D1551354

JAWA

JAWA

LOST CITY OF THE BLACK DESERT

S.W. HELMS

CORNELL UNIVERSITY PRESS
ITHACA, NEW YORK

First published 1981 by Cornell University Press

International Standard Book Number 0-8014-1364-8

Library of Congress Catalog Card Number 80-69820

Printed in the United States of America

For Jordan and its people

Walking in the torrent bed at Ma'an my eyes lighted upon, and I took up, moved and astonished, one after another, seven flints, chipped to an edge; we must suppose them of rational, that is an human labour. But what was that old human kindred which inhabited the land so long before the Semitic race?

Doughty, *Arabia Deserta*

CONTENTS

PREFACE

I would like to dedicate this book to the people and land of Jordan as a study of prehistoric transmigrant urbanism in the desert, over 5000 years ago: to acknowledge Jordan's traditional hospitality and generosity extended to the excavations at Jawa throughout five seasons of work. I am particularly grateful to His Royal Highness Crown Prince Hassan for his gracious interest in the fortunes of Jawa and all of us who worked there.

I have great pleasure in acknowledging the help and support of many Jordanian institutions, the authorities of which always provided invaluable assistance in many practical and theoretical ways: the Department of Antiquities, the Natural Resources Authority, the Jordanian Army and the Royal Scientific Society.

Much timely and useful advice and assistance was also offered by HM British Embassy and the British Council in Amman. H. G. Balfour-Paul, the then British Ambassador to Jordan, always encouraged the project, particularly in its early fragile stages. Grateful thanks must also be extended to Mr Lovett-Turner and Mr Munro of the British Council for their hospitality and moral support.

We could not have worked at Jawa without the sponsorship of the British School of Archaeology in Jerusalem and no more than a sketch-plan of this important site would have been produced had we not been supported financially by the British Academy, the British School of Archaeology, the British Museum, the Seven Pillars of Wisdom Trust, the Ashmolean Museum, the City of Birmingham Museum and Art Gallery, the Manchester Museum, the Central Research Funds Committee of London University and the Palestine Exploration Fund.

Along with the supporters, sponsors and consultants I personally thank warmly all of those who came at their own expense to work at

Jawa, to live some of the time no better than the ancient people whose works we excavated.

Amman, Jordan June, 1979

JAWA EXCAVATIONS. STAFF LIST 1972–6

1972–3 L-A. Hunt, A. G. Walls, H. Qandil, A. J. Amr

1974 H. Ta'ani and A. Saleh (Department of Antiquities), C. Dobson, D. N. Fenner, D. Fleming, I. Furlong, A. Hilton, J. Horner, L-A. Hunt, N. C. Joyce, L. MacLaurin, M. O'Connell, G. Troilett

1975 H. Ta'ani and H. Jasir (Department of Antiquities), R. L. Chapman, C. Davey, C. Dobson, D. Fleming, L-A. Hunt, C. Langdon, M. Macdonald, N. Roberts, C. Rogers, R. Wade

1976 H. Ta'ani (Department of Antiquities), G. L. Harding, R. Duckworth, D. Fleming, M. Macdonald

Photographs, unless otherwise captioned, are by David Fleming. Much of the photographic work in London–preparation of maps, plans, etc. from aerial photography – was made possible through the co-operation and kindness of Mr P. Dorrell of the Institute of Archaeology, London University.

FIGURES

PLATES

between pages 142 and 143

Photographs:
JG John Grey (Qa'a Mejalla survey, 1979)
LAH Lucy-Anne Hunt
SWH S.W. Helms: 1975 aerial reconnaissance by Royal Jordanian Army helicopter.

TABLES

I

JAWA

Father Poidebard's bi-plane over the Black Desert in the 1930s . . . the first photographic record of Jawa

I

THE ANTIQUITY OF JAWA

Thirteen years after the First World War a two-seater bi-plane crossed over a bleak and desolated place called Hosn Jawa, circled several times and then flew on. The French explorer Poidebard had strayed into ostensibly British territory, the mandate of Transjordan, to become the first western man to see this site that was not to be understood until over forty years later.

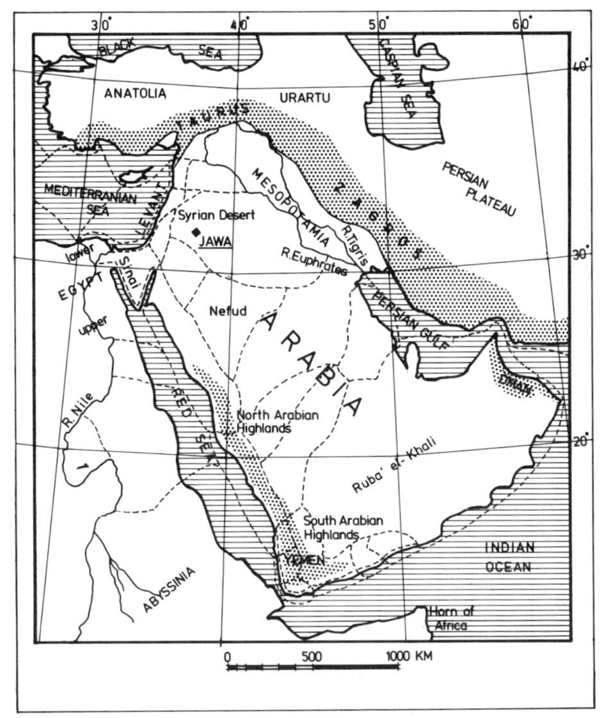

Fig. 1 Map: Middle East trade routes

For several months Poidebard had been surveying and photographing the basalt-strewn desert about Jebel Druze that stretches south from Damascus well into Saudi Arabia. He had crossed a land that another flier had recently described thus: 'except for short periods in the spring the whole of this country looks like a dead fire – nothing but cold ashes'. These are the words of Group Captain Rees who travelled the airmail route from Cairo to Baghdad during the 1920s. He added to this sinister image the possibility that 'it [the land] supported a large permanent population in some period or periods' (Rees 1929). Poidebard had discovered the only large settlement known in this desert that could and did shelter such a population. He did not realize that Jawa was over 5000 years old.

In the 1930s the ruins of Jawa looked much as they do today, and when in 1975 I circled over the site like Poidebard there was only one slight difference: my aircraft was a Jordanian Army scout helicopter obtained through His Royal Highness Crown Prince Hassan. The same strange orderly arrangement of straight walls, at least one gateway, the dam in the west, crater-like pools and the enigmatic web of apparently random walls on the flat land west of Jawa lay silent and grey and so much part of the landscape that it was hard to believe that the site itself was more than a natural formation. Had Poidebard landed at the site he might have realized not only its significance but also its age. But he flew on, attributing Jawa to the Romans, along with all the other places he had surveyed from the Euphrates almost to the Jordan river, summarizing them as 'la trace de Rome dans le désert de Syrie' (Poidebard 1934).

But who did build this 'Lost City' in such a forbidding land of nothing but cold ashes? Who, why and when?

The question of Jawa's antiquity continued to worry us even after it had been answered. It was a lack of confidence in cold facts that would not entirely disappear simply because it seemed both unbelievable and impossibly lucky to find such a place. All who worked there continually buttressed their proofs with a kind of chronological, historical process of elimination that began with Poidebard's Roman camp. What were the Romans doing at Jawa, and why are there no records of a heavily fortified Roman town covering an area of over 100,000 m^2? Such large towns are known to have existed through records as well as surveys and excavations, but there is no mention of anything at Jawa. Purely on these grounds it appeared that our town must be earlier than the Roman period; that is, prior to Pompey's annexation of Seleucia in 64 BC. The same applied to any of the other political entities governing the area of Jawa as a marginal region attached to more verdant lands: the Seleucid heirs of Alexander the Great, for example, or the Nabateans of Petra.

Turning to more empirical methods it should be noted that the masonry at Jawa is very different from that of settlements further west and north (Umm el-Jimal, Bosra or Salkhad), being made from unworked stone set in rough courses. Jawa's architecture may be as massive but it is not so carefully finished; perhaps because the requisite tools were not then available. There is only one architectural feature at Jawa that could possibly be attributed to a later period and that occurs in the large rectangular building complex visible at the summit of the site. This so-called Citadel has a roof made from cantilevered basalt beams, a technique that is well known at Nabatean and Roman sites throughout Jordan and Syria. But for elimination dating here this is of no real consequence since the Citadel is obviously later than the town whose hut-circles pass underneath it.

The next clue comes from several building blocks in the fortifications. Some of these bear pre-Islamic inscriptions and animal carvings of the Safaitic bedouin who lived in the area from the first century BC onwards – at least that is the earliest record of their 'literacy', for their identity other than a general nomadic one was as

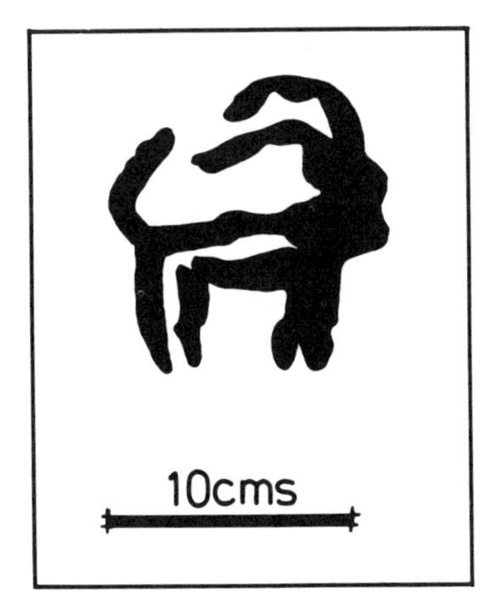

Fig. 2 Safaitic animal drawing at Jawa

much a puzzle as Jawa. So far as dating the town is concerned, short of tearing up the whole site the probability must be conceded that these messages and pictures were carved after the place had been built. Again one is left with a pre-Roman date. One is left with a Jawa, moreover, that cannot be attributed to the only known folk of this desert, the Safaitic beduin; and this is supported by the pottery sherds scattered all over the site. Even without knowing their true identity we know what these sherds are not. They have nothing whatever to do with pre-Islamic, Roman or Nabatean wares, which are unmistakable.

So one is inexorably pushed back through time, past Alexander's conquest to the period of the Persian suzerainty over the land that would be this part of Arabia (538–332 BC). Once again there is silence. Similarly, when one thinks in terms of Babylonian rule that began in the early sixth century BC, there are no records of any city east of Bashan (Jebel Druze) in this region and no specific mention of a local kingdom under whose sway such a large and therefore important town might have been constructed. Thus it goes on until by the process of elimination even the Age of the Patriarchs – whenever that was – is passed and one is left with vague textual references in biblical genealogical tables: 'there are the sons of Ishmael, and there are their names, by their towns and by their castles; twelve princes according to their nation' (Genesis 25:16) – Ishmaelites being the ancestors of the Arabs, including the Safaitic beduin.

By this point, having arrived at some date before the second millennium BC, there would be a strong suspicion that Jawa is very ancient indeed; no matter how extraordinary the implications. And at this stage pottery first entered the chain of reasoning and the meagre yield from a quick survey sondage in the Citadel (1972) produced a date between about 2000 and 1800 BC for that building which, as has been said, is much later than the town. Through elimination a date had been arrived at for Jawa of the third millennium BC, when the first cities appeared in Palestine to the west; and even this was too recent as it turned out.

Table 1 Archaeological Periods

*Date**	*Egypt*	*Palestine*	*Syria/Mesopotamia*
		PALEOLITHIC MESOLITHIC (from *c.* 10,000 BC)	
		NEOLITHIC (ACERAMIC/CERAMIC) (*c.* 8th–5th millennium)	HALAF PERIOD
	FAIYUM 'A' NEOLITHIC	CHALCOLITHIC (GHASSULIAN)	SAMARRA/HASSUNA
3500	PREDYNASTIC PERIOD	(*c.* 5th–4th millennium)	UBAID PERIOD
	PROTODYNASTIC PERIOD	LATE CHALCOLITHIC or EB1 (A, B, C) or PROTO URBAN (A, B, C) — villages only, immigration from north and east	trend towards urbanism URUK VI–IV
			JEMDET NASR (AMUQ G)
	EARLY DYNASTIC PERIOD		
3000	dynasties 1–2	EARLY BRONZE AGE 2 heavily fortified towns and municipal water supply systems	EARLY DYNASTIC PERIOD NINEVITE 5
	OLD KINGDOM	EARLY BRONZE AGE 3	
	dynasties 3–6	apparent increase in planning	proto-imperial urbanism
2500	(GREAT PYRAMIDS)	control of towns, greater uniformity in fortification design	
			(EBLA) SARGON OF AKKAD
		EARLY BRONZE AGE 4 MIDDLE BRONZE AGE 1	
	1st INTERMEDIATE PERIOD	?non-urban?	NEO-SUMERIAN AND OLD
2000	dynasties 7–11	immigration	BABYLONIAN PERIODS
	MIDDLE KINGDOM		Ur
		MIDDLE BRONZE AGE 2 massive urban fortifications	Larsa Isin
	2nd INTERMEDIATE PERIOD dynasties 13–17	and design uniformity throughout the Near East: urban water supply systems increased in scale and efficiency	Babylon
1500			

**approximate*

2

JAWA'S LEGACY?

We had come close to the truth, just as the first explorer to reach Jawa twenty years after Poidebard, Gerald Lancaster Harding, then Director General of Antiquities in Jordan, who dated the site in the Early Bronze Age. Excavations were to show that Jawa was a little earlier than this and was built during the latter part of the fourth millennium BC. We realized (hence our initial scepticism) that we had before us the earliest and largest developed town in Jordan.

Jawa is the best preserved fourth-millennium town yet discovered anywhere in the world: paradoxically in a place – the Black Desert – where it could hardly exist today and probably hardly when it was built. It turned out (another paradox) that the town only lasted for a generation or even much less, being created virtually by accident and thus so wonderfully preserved.

Once the date is accepted another even greater idea has to be faced. Jawa may represent a rare example of measurable, direct demographic links through technological precedent and the diffusion of ideas. The urban consciousness and technology of nearby Palestine follows almost immediately on the abandonment of Jawa. Fortifications apart, the one aspect of the site that is an essential part of permanent and secure settlement even in verdant regions concerns the problem of water supply. No walled city survives long without an internal water-storage system: this is a fact of siegecraft which is an urban reality. Unlike any other archaeological site dating from near the beginning of cities in the Near East, Jawa represents an urban hydro-technology that is almost completely preserved above the ground and therefore measurable. This prehistoric scientific consciousness and the systems it produced might be compared to what occurred in Palestine during the Early Bronze Age shortly after the end of Jawa.

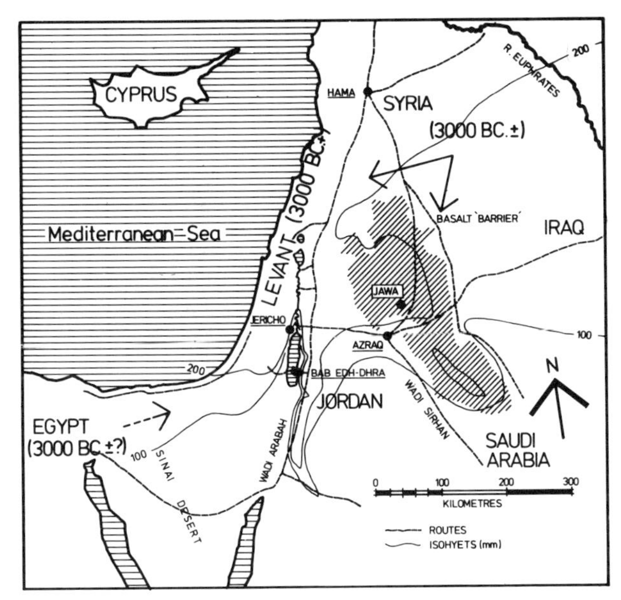

Fig. 3 Map: beginning of urbanism

The establishment of the town and its essential life-support systems at Jawa took place during a time of well-attested population movements throughout the greater region and might be seen as a part of this pattern. The technical excellence of the support systems and the high level of the kind of science inherent in them, as well as the geographical location of the site, lead to the hypothesis that the people of Jawa came from a developed urban tradition. Since this prehistoric technical brilliance only lasted a short time, and so near to the gentler and relatively uncontested lands of the Levant to the west, Jawa is truly a paradox in the history of urbanism that has nevertheless preserved a potential recognition of prehistoric ideas – science among them. Ironically, preservation is due to the nature of this harsh land in which Jawa was built. After the fourth millennium no town was ever built there again.

Even a brief examination of Early Bronze Age Palestinian water systems – schematized here – demonstrates at least a technological connection with Jawa. It is now understood that towns in the more arid regions relied on surface water resulting from winter rains which was collected and stored in man-made reservoirs or cisterns. It will be seen in due course that all of Jawa's water systems

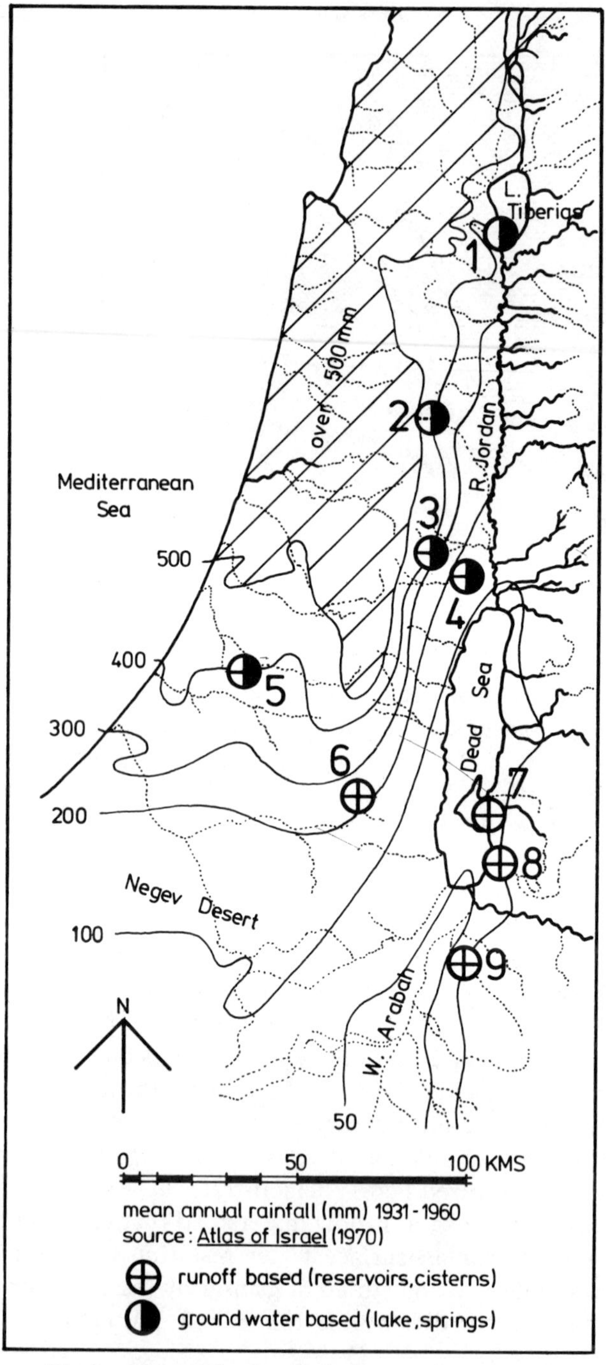

Fig. 4 Map: Palestine Early Bronze Age towns: urban water supply (*c.* 3000 BC+)

depended on such resources. Thus in Palestine, Tell Arad apparently relied totally on winter rains that were harnessed according to what might be called the Jawa method from about 3000 BC onwards. Somewhat similar systems may have existed at the town of Bab edh-Dhra' above the Dead Sea in Jordan (deflection of water into canals) and at Tell 'Areyny in south-western Israel. Even at the town of 'Ai (et-Tell), which lies in the more verdant Judean Hills north of Jerusalem, surface water was collected in at least one pool or cistern to supplement the perennial source beyond the walls. With the exception of Beth Yerah (Khirbet Kerak), which lies on the shore of Lake Tiberias, many of the new urban Palestinian water systems may have used earth and stone structures very similar to the dams of Jawa. At Jericho, for example, in the otherwise semi-arid southern Jordan Valley, such structures could have contained water from the abundant spring of 'Ain es-Sultan as a convenient and in times of siege crucial storage system.

Thus arises the hypothesis that the technocrats of late fourth-millennium Jawa may easily have emigrated yet another time and entered the Land of Milk and Honey along with other migrants and been the source of the new urban idea in the land that became known as Canaan. So the work at Jawa is of the utmost importance, not only as a prehistoric legacy of man's achievement but also as of primary relevance to modern man, who after all still lives in cities with disturbingly similar problems.

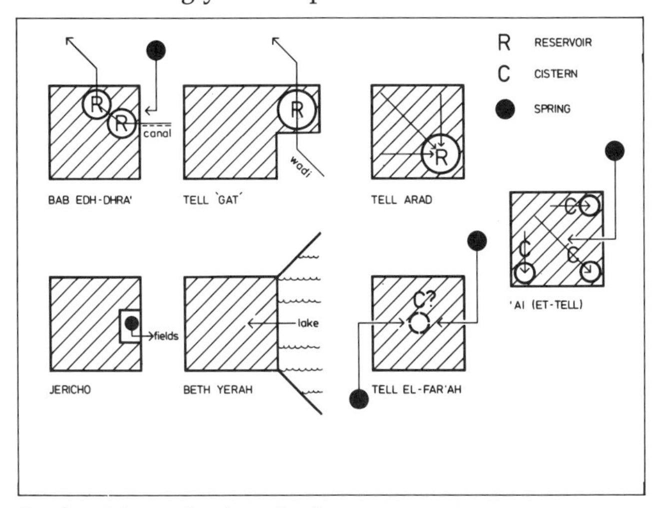

Fig. 5 Schematic plan of urban water systems after *c.* 3000 BC

II

THE BLACK DESERT AND ITS PEOPLE

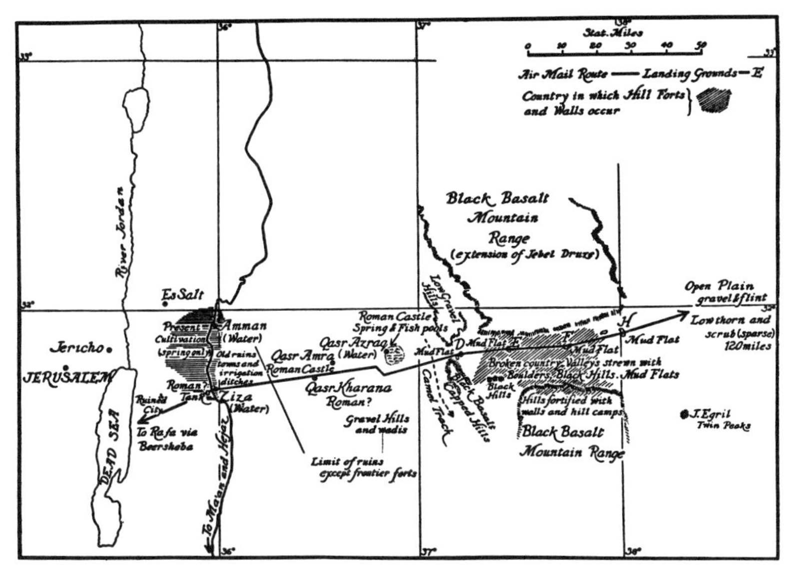

Flight Lieutenant Maitland's map: the airmail route from Cairo to Baghdad along which ex-RAF pilots discovered the mysterious 'desert kites' and other enigmatic signs of man's past in the Black Desert

3

DISCOVERY

Jawa is a paradox rife with irony that even colours the process of its discovery. Ever since the nineteenth century and the dawn of Arabian exploration, many people have come close to discovering the site. Euting and Huber passed through the Jebel Druze area west of Jawa on their way south. Lady Anne Blunt even travelled along the wadi Rajil – Jawa's 'lifeline', so to speak – and among her many sketches is one depicting a sandstorm in that very place, a poignant record of this desert's harshness. During the first two decades of the present century a well-organized expedition under Butler from Princeton University reached the ruins of Deir el-Kinn just 15 km north-west of Jawa, but went no farther. Sir Aurel Stein, Gertrude Bell, Henry Field and others approached from the east. But none found Jawa, hidden in the black hills at the foot of Jebel Druze. Had they done so they would surely have written about such an enigma in the desert – if only as a vignette among the many mysteries of Arabia that they sought.

Perhaps the best example of Jawa's persistent elusiveness is a lightning survey by Nelson Glueck just before 1948. He crossed the Black Desert as far as Qasr Burqu' – a lonely outpost on the eastern boundary of the Roman world – moving fast and seeing little because of inclement weather, impending political troubles in the west and, no doubt, a measure of that reluctance to tarry long in this wasteland that is apparent from all such explorations.

Glueck must have passed within 200 m of the western fortifications of Jawa. He saw nothing, writing later: 'Jawa marks the location of a small, filthy spring, below the west end of which a crude birkeh (pool), about 20 m in circumference, has been dug. . . . There was probably never more than a small police post at Jawa' (Glueck 1951). A filthy spring, indeed. Did he not know about

Poidebard's aerial photograph published nearly twenty years before? Could he not have stopped long enough to ask the bedouin at Deir el-Kahf or Deir el-Kinn – where he did stop – about an ancient site with an area of more than 100,000 m^2.

The irony compounds itself. Just one year before Glueck's published judgement of Jawa an epigraphical expedition reached the site and described a large black fortified town with a huge dam (Winnett 1951). Thus while one party was dismissing Jawa as a 'filthy' spring and no more, another had at long last reached the real place that Poidebard had seen from his bi-plane in the 1930s. But still nothing was done about Jawa other than collecting a few hundred pre-Islamic inscriptions: that is, almost nothing. One of the investigators, Harding, returned to Amman that year and wrote in the records of the Department of Antiquities the words that I was to see in 1972: 'Jawa: Early Bronze Age?'

Since the 1950s Harding had shared his enthusiasm over Jawa and told his colleagues about this strange ruin. He was still doing this when I arrived in Jerusalem in 1966. People were mildly intrigued, but loath to investigate, presumably since so much else was being found in Jordan and Palestine in more accessible, more congenial places. By 1966, however, many parts of these lands had been allotted to various spheres of archaeological interest, leaving only the nastier sections unclaimed. Out of no more elevated motive than going where none would go, I went east late that year and finally reached Jawa on 14 November – becoming the second, albeit solo, 'expedition' to see the place since 1950.

Further work was interrupted by an event that was essentially the modern expression of a timeless pattern in the Near East: the Arab-Israeli war of 1967. Again it is an irony, because this is the very pattern of which Jawa is the earliest measurable example.

In 1972 I revived Jawa and conducted excavations there until 1975–6. At long last this enigmatic silent ruin was truly found; the mysterious phantom of Poidebard's survey that everyone had either missed or misinterpreted, and the Black Desert that all were either unwilling to enter or, once there, eager to quit at the first opportunity.

4

THE FORMATION OF THE LAND

The bedouin call the land around Jawa Harrat er-Rajil. Harra means a stony and volcanic area, a lava field; Rajil refers to the major valley beside Jawa, the word itself signifying walking men. In their words then the Black Desert is the stony land of the walking men, the land of volcanoes which was made long ago and the work of the Devil.

It is bleak and forbidding, almost a true desert precisely because of its formation and subterranean origins. A land covering over 45,000 km^2 of which over 11,000 km^2 lie within the Hashemite Kingdom of Jordan (itself only 96,500 km^2), this desert has been called the Basalt Barrier (Baly and Tushingham 1971) and stretches from just south of Damascus (the Leja and Safa) south to Jauf at the northern edge of the infamous Nafudh Desert of Saudi Arabia. The basalt lies directly across the shortest route between the Euphrates river of Mesopotamia and the Jordan Valley. Even the cold scientific terms describing this land, like the bedouin names, conjure up a kind of hell on earth that has led almost all who have travelled there to evoke Dante, Milton or anyone who ever referred to the works of Satan. In sympathy with bedouin superstition one of our representatives from the Department of Antiquities applied the most recent and very apposite epithet, calling the area Bilad esh-Shaytan, the Land of the Devil.

The region is part of the Arabian Peninsula and is surrounded by the interior deserts that lie north of the line from the Gulf of Aqaba through Jauf to Kuweit on the Persian Gulf, the area called the Badiat esh-Sham or Steppes of the North which, but for the basalt extrusions, consists of relatively open desert, part steppe. To the south is the Nafudh Desert, the scene of T. E. Lawrence's many trials. On the west lies the land called Ardh es-Sawwan, the Land of Flint, and on the east the Hamada and Wudian. The latter are a broad series of more or less parallel wadis that tend north-east through open rolling country in which travel is relatively easy. This is an area of much nomadic activity where even today the Ruwalla bedouin move each year, without much regard for international boundaries, from the Turkish frontier to the southern tip of Arabia.

From Jauf in the south the interior of Arabia is linked with Jordan and Syria by wadi Sirhan which skirts the western shore of the basalt, deriving from it and the western Ardh es-Sawwan plentiful water from winter rains. There are numerous wells along this much-travelled trade route that ends at the oasis of el-Azraq at the edge of the basalt just under 60 km south-west of Jawa. Wadi Rajil is one of the many ephemeral streams that combine to make el-Azraq a rich oasis: Azraq in Arabic means blue.

To the north of the basalt region lies a long north-east tending chain of hills called the Jibal esh-Sharq, or Mountains of the East. They form the southern boundary of the good Syrian Steppe that reaches north to the great arc of the mountains framing the Arabian Peninsula as well as the Mesopotamian Trough and the Levant. The verdant parts there are called the Fertile Crescent. Our region, the greater Harrat er-Rajil or Bilad esh-Shaytan, the Black Desert, lies like an island in a sea of deserts at the apex of Arabia. There, almost at its centre and under the shadow of an extinct volcano, lies the ruined town of Jawa.

The geological history of the greater region, the northern end of the so-called Fertile Crescent, is relatively simple. It is bounded by the arc of the young Anatolian and Iranian fold mountains in the north and in the south by the stable shelf of Arabia flanked by the Nubo-Arabian Shelf to the west. Between this and the mountains lies the unstable shelf of the Tethys Ocean which once joined what is now the Mediterranean Sea and the Indian Ocean. This was about 20,000,000 years ago and great amounts of marine material were deposited then whose fossil remains can still be found. It is an interesting if peripheral thought that the land on which Jawa was to become a milestone of ancient hydro-technology once lay on the shore of a primeval ocean.

The Black Desert lies on the east-Jordanian limestone plateau that stretches from the rim of wadi Araba and the Jordan Rift Valley east to Iraq. It is made of basalt and tuff from Jebel Druze, the Uneiza and Ruwalla highlands and covered by the extensive lava fields that

Table 2 Geological Periods

Era	*Period*	*Epoch*	*Duration*	*Years ago*
Cenozoic	Quaternary	Holocene	*c.* last ·01	
		Pleistocene	2·5	2·5
	Tertiary	Pliocene	4·5	7
		Miocene	19	26
		Oligocene	12	38
		Eocene	16	54
		Paleocene	11	65
Mesozoic	Cretaceous		71	136
	Jurassic		54	190
	Triassic		35	225
Paleozoic				

(years in millions)

begin just east of the modern town of Mafraq. The area is occupied mainly by Upper Cretaceous and Tertiary Calcareous sediments: a very recent chapter of creation. Jebel Druze is the mother of this amazing land of volcanoes along the Sirhan Depression and a long line of fissure effusions, chains of smaller volcanoes, maars and tuff-pipes. No wonder the name Bilad esh-Shaytan.

This land on which Jawa was to be built is as complicated as it is forbidding. There are basalt sheets, basalt and tuff volcanoes and fissure effusions of the Jebel Druze area north of el-Azraq; basalt and tuff volcanoes and maars at the southern rim of Jebel Druze; basalt sheets and crypto-volcanic structures south of el-Azraq. 50–100 km wide, the lava flowed west and south-east of the Azraq Depression. The lava of Jebel Uneiza and Ruwalla (al-Harra) flowed into what is now Saudi Arabia for 210 km along wadi Sirhan. The basalt region is therefore nearly half the size of Jordan and larger than Palestine. It is a mad cast-sculpture representing an enormous amount of basalt lava from numerous fissure effusions, some still visible as long rows of cones following fissure lines and even off these lines basalts with some tuff were extruded from clusters of isolated eruption cones. As we worked in this wasteland, especially while moving through it on

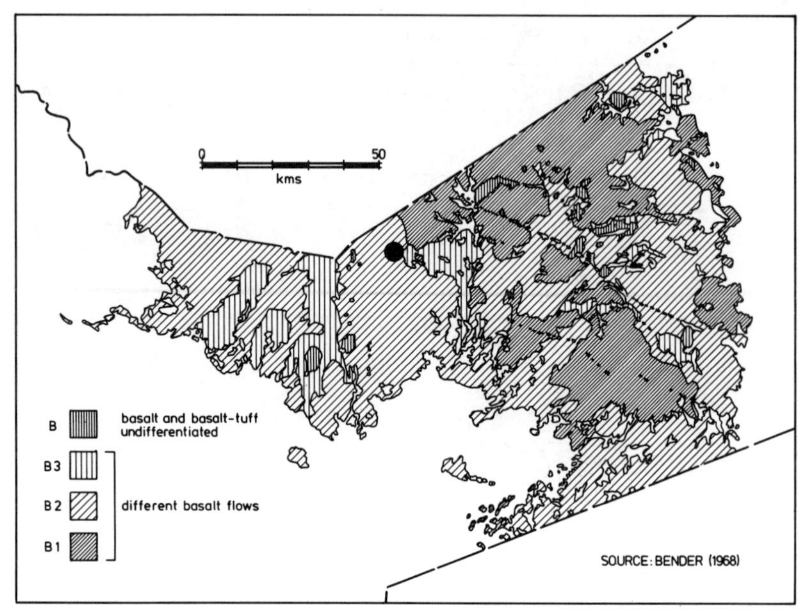

Fig. 6 Map: Pleistocene and Neogene basalts in NE Jordan

surveys, we were often reminded of fairy-tale landscapes and, most poignantly, of creation myths evoked by the thought of how this land about Jawa must have been formed out of a seething chaotic mass of molten rock.

This fundamental part of the Jawa story, the formation of the land, took place relatively recently. Geologists have recognized six major emissions in the region that caused what we see today. The lowest and earliest of these is not well documented since exploration is still in its infancy. Evidence comes from limited drill samples in wadi Dhuleil.

The first three emissions produced lava over 150 m thick overlying Middle and Upper Eocene limestones. The three flows are separated from each other by fossil soils about 5 m thick, red clays and other weathered surfaces. The uppermost flow is overlaid by 6–20 m of various soils which in turn are covered with the fourth flow of about 60 m thickness. This in turn is covered by sandstone, calcareous sandstones and sandy marls of the Miocene age. Thus the first four flows started at the earliest in the Upper Eocene, after 54,000,000 years ago, and ended in the Miocene between 26,000,000 and 7,000,000 years ago.

The fifth flow forms the major part of the exposed basalt in the north-east of Jordan today. It is up to 25 m thick and in places even thicker, forming cones of irregular relief on top of older basalts with

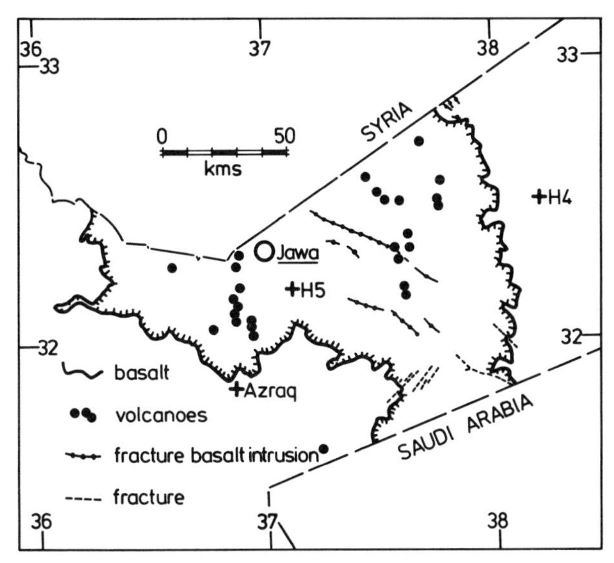

Fig. 7 Map: the Black Desert volcanoes

various soil deposits, clay horizons and tuffs, all on top of Paleocene sediments and Miocene calcareous sandstones, thus dating the flow in the interval between the Miocene and Pleistocene.

After this fifth flow come the main tuff eruptions, overlying it as fine and coarse tuffs, either as small occurrences of thin sheets or as tuff volcanoes. The centres of these tuff eruptions lie 25–35 km west of Jebel Aritain and 50–70 km north-east of the pumping station H5.

The sixth emission produced basalt flows that spread over several kilometres and their cooled surfaces are virtually unweathered today. This flow overlies the older basalts and their weathering and occasionally also the tuff of Jebel Aritain south of Jawa. The flows are more than 10 km wide and up to 30 m thick. They cross the Mafraq to Baghdad road and can be dated according to certain archaeological evidence in the Middle Pleistocene Epoch which begins about 2,500,000 years ago.

Finally there are fissure effusions kilometres long that are said to belong to the most recent eruptions in the region on stratigraphical grounds. About 48 km east-north-east of the pumping station H5 is a row of such volcanoes 90 km long. Others follow fissure zones to the north-east (20 km long), east-south-east (60 km long) and 55 km south-east (35 km long). All strike the direction between 110° and 130°. Bender in his *Geology of Jordan* (1974) notes:

it is possible that the latest eruptions have continued to historical times and correlate with the youngest lavas of the Jebel Druze in Syria. . . . Organic matter found in a basalt there had (sic) been determined by the C^{14} method and resulted in an age of about 4000 years.

The formation of the Black Desert was still taking place long after the town had been abandoned.

Beneath this complicated volcanic cover the structure of the region and those surrounding it is summarized here (figure 8a) in simplified form: (1) underlying structures obscured by recent sediments; (2) sedimentary rock in relatively level strata; (3) sedimentary rocks moderately disturbed; (4) archaean platform or peneplated fold mountains; (5) young fold mountains, the Zagros and Taurus ranges.

Climatically the region of Jawa is on the boundary between the types called Saharan-Arabian (1) and Syrian Steppe (2) (figure 8b). In terms of Köppen's classification this is represented by the symbols BWhs and BShs where B is dry climate, BS steppe, BW desert, s dry season in summer, and h dry and hot with mean annual temperatures over 10° Cent. (64° Fahr.). According to the moisture scale (Thornthwaite: regions E and D) this area is semi-arid to arid with annual temperatures as noted and common occurrences of temperatures well above 38° Cent. (100° Fahr.). It is a region of great daily and annual temperature ranges with rain only during the winter months, a belt north of the tropical deserts of North Africa and Arabia with conditions that also occur in the north-east section of the west Arabian Highlands and the mountains of Oman.

Vegetation (figure 8c) today is typical of Plains Steppe (4), the Jawa area specifically being semi-desertic (3) bounded by the Hamada (2) or stony desert to the east and, further south, Erg or sand-dune desert, the Nafudh (1). It is a land of transition between near-desertic vegetation and poor steppic. Plants are suited to dry conditions and often to salty soil. The region abounds in thorny

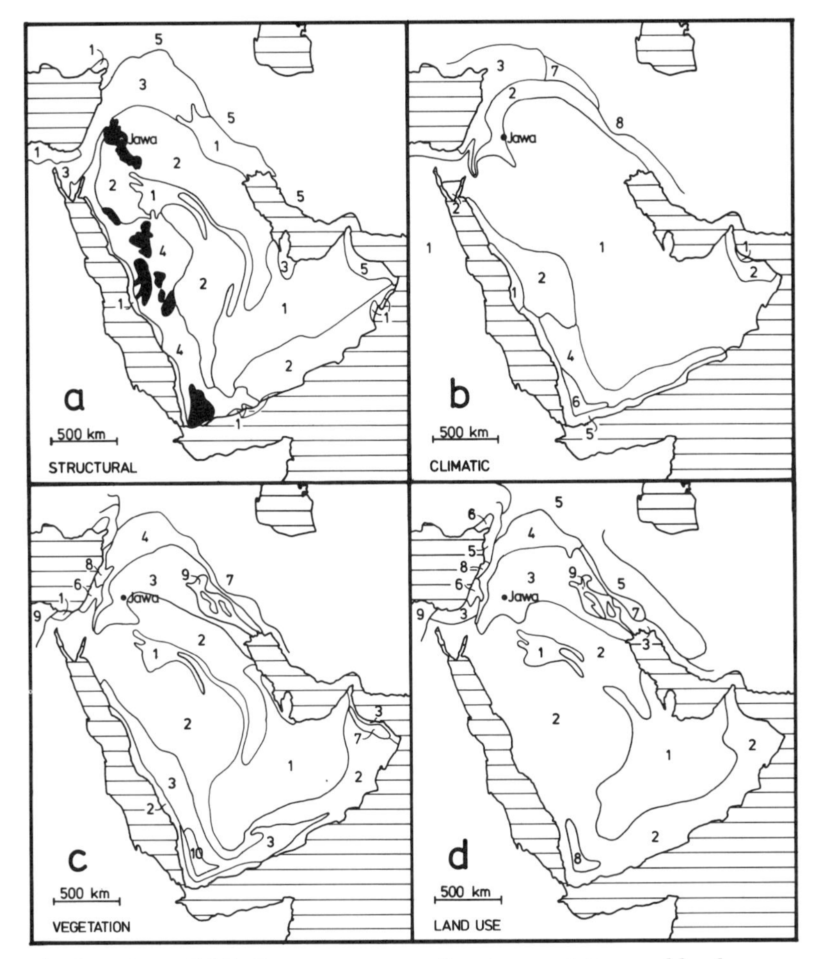

Fig. 8 Map: Middle East – structure, climate, vegetation and land-use

bushes and shrubs such as tamarisk. Vegetation flourishes in wadi beds and on the north side of hills. In the spring, and even into the summer after a wet winter, the desert here explodes and makes itself bloom quite without the help of man in an unbelievable variety of plants, many of which are still being recorded and classified. They include hawthorn bushes and lavender as well as all cures for ailments ranging from dysentery to unwanted pregnancy. In the Jebel Druze area oak is still found.

The predominant form of land-use (figure 8d) is pastoralism, mixed camel and sheep/goat and hardly any agriculture so far as general classification of the region is concerned. Thus barley cultivation, though possible in the dryer parts, is relegated to the northern areas where it gradually gives way to wheat as conditions become wetter. The basalt, however, is a special sub-region in this regard as in many others. In addition to this there are two special areas: the oases on the edge of the basalt, el-Azraq and Qasr Burqu'. The Black Desert is a marginal zone and in the recent past was the interface of the free-roaming camel herds and the settled land. In times of peace the shepherds from the villages to the west will take their flocks far into this desert during the summer for what grazing there might be. On the other hand the bedouin do practise a form of casual agriculture (random seeding), and visit their 'fields' only at harvest time in the hope that some rain has fallen on them. It is a margin, this Black Desert, across which the timeless pattern of settlement and its opposite – banishment and enforced return to a more basic form of life – takes place. Today many bedouin are settling down to a kind of true mixed farming.

This then is the character and most ancient history of the land where Jawa was built. The formation of its topography determined settlement types and patterns. After the lava had cooled and set, the annual rhythm of rain falling on higher ground and draining over the basalt and tuffs cut the valleys that ultimately took the shape so suitable for man's survival – should any man choose to live there. The land was sculpted into valleys and places that filled with erosional material and aeolian deposits to form the typical mudflats that are the only visual relief in this black land. The mudflats were to play a part in the life-support systems of the prehistoric period and still supply the locally grown food for the semi-settled bedouin.

The soils generated within the desert derive from weathered basalts, but much of what is seen in the mudflats and also at Jawa is transported by the wind. This includes steppe soils from Syria and west Jordan, the parent material being composed of loess-like sediments, calcareous rocks and basalts as well as sandstone. These soils – in the area of Mafraq and Jiza – are especially used for the production of barley and pasture land. Their agricultural yield can be improved by intensive irrigation. Also wind-transported into the basalt region are the grey desert soils (sierosem) which develop in areas of less than 150 mm of annual precipitation and occupy more than half of Jordan, specifically surrounding the Black Desert on all sides. Without irrigation this type of soil provides only poor grazing.

In spite of all this bleakness the Black Desert can and has in the past supported life at a tolerable level. The reason of course is simply water, and for Jawa the special hydrological situation created by the formation of the land. This aspect, the archaeo-hydrology, is discussed in Section V. Here one need only recognize that because of the structural history of the region there is no readily exploitable groundwater, except for the lakes at el-Azraq and Qasr Burqu' which of course lie outside the area. The water is too deeply buried beneath the frozen lava. Within the Black Desert the only water that can support life comes from the winter rains, from water that remains on the surface of the ground in an ephemeral way as surface run-off and drains southwards along a series of wadi systems of which one, wadi Rajil, passes by Jawa. Water is held momentarily in the mudflats, but much that does not evaporate or is absorbed by plants ultimately reaches the oases outside the basalt cover. In our area water always moves rapidly during very short periods of time. It is elusive, sporadic and unpredictable like the wild life of the desert that must first be understood and then harnessed, herded and corralled before it can support man.

5

HISTORY AND PREHISTORY

The desert of Jawa enters history sporadically as a frontier region and any permanent building activity within it is intrusive. But in addition to the greater historical pattern that typifies this and any other desert as a frontier there is the fact of an internal, indigenous history that may not be of a literate folk but is nevertheless a constant presence, based on the life-style and subsistence pattern dictated by the physical environment. Empires and kingdoms passed; the nomads of the desert continued throughout, before during and after such intrusions, like their cousins the fellaheen in the historical lands who also tend to survive changes of government. If they do not, they become refugees and for a time at least, and in some cases perhaps once again, nomads.

For these reasons, quite understandably, the human history of the basalt desert is not complete. It is intermittent, as implied by a frontier region, and sparse by the very nature of nomadic life that leaves few signs of its passing.

Man lived or at least passed through the region during the Mesolithic period, after about 10,000 BC (confining ourselves to post-glacial times), before agriculture was evolved. He probably lived by hunting and a little later by some form of pastoralism. His meagre traces have been found in the form of stone tools scattered near caves and the ephemeral water sources in wadi Rajil just north of Jawa. As yet little more than that is known.

Until very recently only vague indications existed of human activity in the basalt desert during the next era and these were confined to the more permanent water sources such as the oasis of el-Azraq. Now, at last, there is sufficient evidence to state categorically that a large permanent nomadic population dwelled there throughout the Neolithic period after about 8000 BC right up to the creation of urban Jawa at the end of the fourth millennium. Moreover it can be assumed that these people were probably no less than the ancestors of the bedouin who are now beginning to settle in permanent homes. So dramatic is this apparently minor discovery, once the full meaning is realized, that for the period between *c.* 8000 BC and Jawa – nearly 4000 years – a separate chapter must be devoted.

There are however a few discoveries at Jawa that cannot yet be firmly set into a continuous prehistory. These concern rock drawings that are definitely earlier than the ubiquitous pre-Islamic Safaitic inscriptions and pictures that were the primary interest of Winnett and Harding in the 1950s. The precise date – even the general period – is not in the least settled and it is doubtful whether it will ever be possible to ascertain it in the strictest stratigraphical sense. This is the case both at Jawa and in Arabia generally where comparable examples have been found.

The best examples representing cattle are drawn on one stone just up the valley from Jawa. We called this gallery of portraits 'Animal Farm'.

Fig. 9 Animal carvings at Jawa: cattle

It is of course a little unexpected to find bulls and cows in the essentially waterless basalt desert. Surely they could only have lived during a period when abundant water was available, which could only have been when Jawa functioned as a town? It is not possible to be entirely certain. The motives of artists are never that clear – even when they themselves describe them, which these folk could not do in any case. These carvers might, for example, have been indulging in wishful thinking; they might have been describing animals that they knew in another place altogether: a kind of rock-art nostalgia

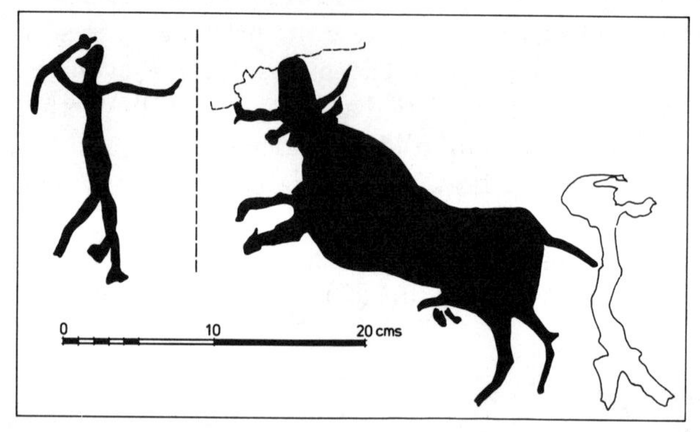

Fig. 10 Carvings at Jawa: man and bull

quite apart from religion. The oasis of Azraq (about 50 km away) still provides grazing for cattle. Or one could look even further afield, albeit via urban Jawa.

The only 'evidence' there is which might link 'Animal Farm' with ancient Jawa comes from the faunal remains which include domesticated cattle (see Appendix E). Domesticated longhorn cattle such as Jawa's are known for the Ubaid period of Mesopotamia (5th–4th millennium) and were introduced into Egypt in predynastic times, appearing on a number of slate palettes, including that of King Narmer. This Egyptian connection will reappear later.

Stylistic and more generally, topical parallels exist for these particular animals in the rock art of Arabia (Anati 1968–72; Adams *et al.* 1977). The connections, however, are even more tenuous than those with urban Jawa; but if they are true it may be possible to say that they are the work of an indigenous people who lived in the basalt regions of Syria, Jordan and Saudi Arabia before and during the time when Jawa was built. This must be the subject of a separate chapter.

Fig. 11 King Narmer and animal pen, Egypt *c.* 3100 BC

Cattle apart, there are several other even more enigmatic rock drawings in the area of Jawa, all of which must be earlier than the Safaitic period. One depicts the hunt of a longhorned beast, the hunter apparently leaping through its horns in the manner of Minoan bull vaulters. This animal, like the cattle, is similar to some Arabian rock art which is rather dubiously attributed to the third millennium BC. Still other Jawa drawings tend towards the

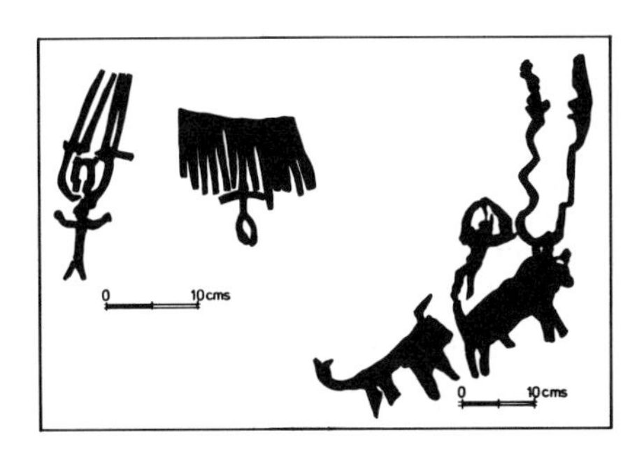

Fig. 12 Carvings at Jawa: thing (?) and a hunting scene

ostensibly mystical and cultic that are quite out of place in this story. They seem to represent either shaman-like creatures with feathery headgear or – as one archaeologist has interpreted similar things – vultures descending on headless corpses left out for that specific purpose (Mellaart 1964). Similar figures are found in Arabia, but as usual without satisfactory dates. The original meaning of these drawings is as nebulous as the necromantic interpretation noted above. The fact that no graves have been found at Jawa probably has no relevance.

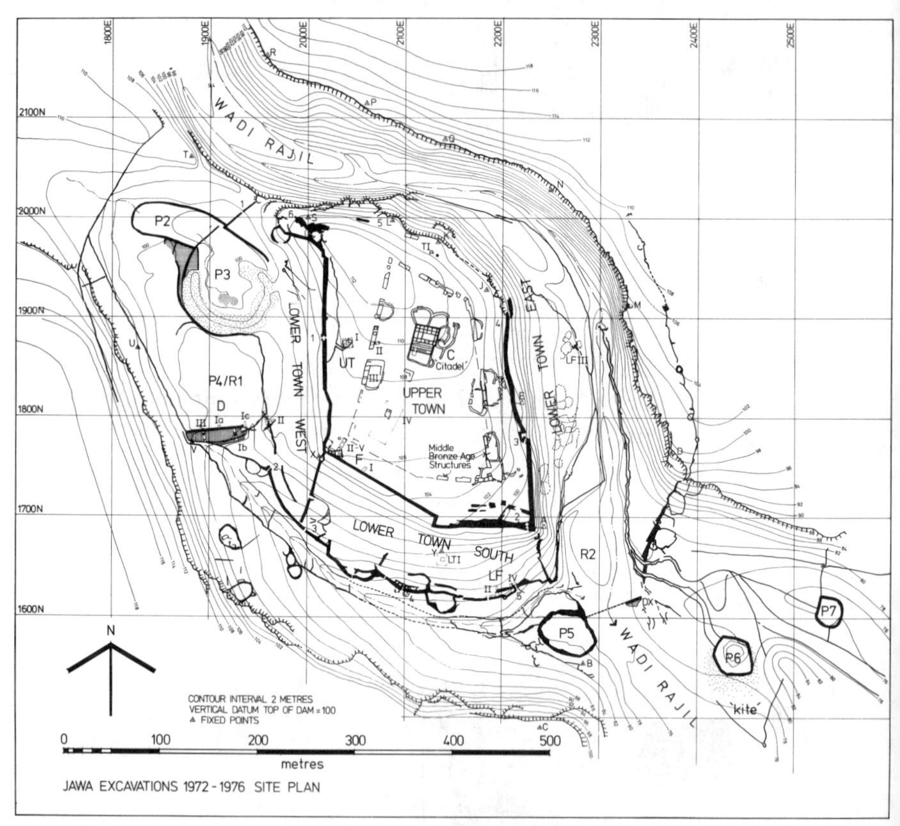

Fig. 13 Jawa site plan and the Middle Bronze Age complex

The Black Desert enters written history with biblical references to Jebel Druze. Referring to much later realities, the mountain of Bashan is described as an area where oak trees grow and where people in disgrace might find refuge in exile (Psalms 68:14–15; Ezekiel 27:5–6; Jeremiah 22:20; Isaiah 33:9). In general this is the region that is part of the posterity of Noah (Genesis 10:21) and the land of the Ishmaelites, descended from Abraham (Genesis 16, etc.), the ancestors of the bedouin, just as it has been intimated above that there may be prehistorical roots that go back to the Neolithic period. This area is also part of later kingdoms (Edom and Amon) and biblical reference is made to adjacent lands: Hauran, Aram, Jaulan and Gaulonites. It belongs to Celosyria of the Hellenistic era and was part of the Seleucid empire after Alexander the Great. Later still it is the land of nomads east of the cities of the Decapolis: Damascus, Raphama, Kanata, Gerasa and Philadelphia (Amman), to name the nearest. But before these times the only archaeological proof of 'history' comes from certain scattered sites dating from the Middle Bronze Age or the Age of the Patriarchs as it has been called, some time after 2000 BC. These sites lie in the basalt desert east and north-east of Jebel Druze.

For the so-called Age of the Patriarchs Jawa has provided some interesting evidence. Once again, because of the location of the site, some speculative generalizations might be made regarding routes of historical, of not biblical significance.

As was noted earlier, the most recent architecture at Jawa belongs to the Middle Bronze Age. At the summit of the most ancient town we discovered a building complex that was the cause of yet more uncomplimentary epithets. Bedouin sometimes refer to Jawa as *deir mahruq* or burned monastery, aptly describing the colour of the ruins; they also call the place 'the prison'. The complex at the top of Jawa consists of a central rectangular building, the Citadel, and several outbuildings describing an irregular pentagon. Each of the outbuildings has a walled courtyard facing the Citadel and that structure is divided into at least twenty-four cells and three transverse corridors. This building which is nearly 4000 years old is so well preserved that most of it is still roofed with cantilevered basalt slabs on piers. Traces of an upper storey also exist allowing us to reconstruct the building as it may have been originally.

How did the people responsible for this complex manage to live at Jawa? What was the purpose of this carefully designed group of structures? And – the usual query – what was the relation of this desert establishment to the world beyond the basalt?

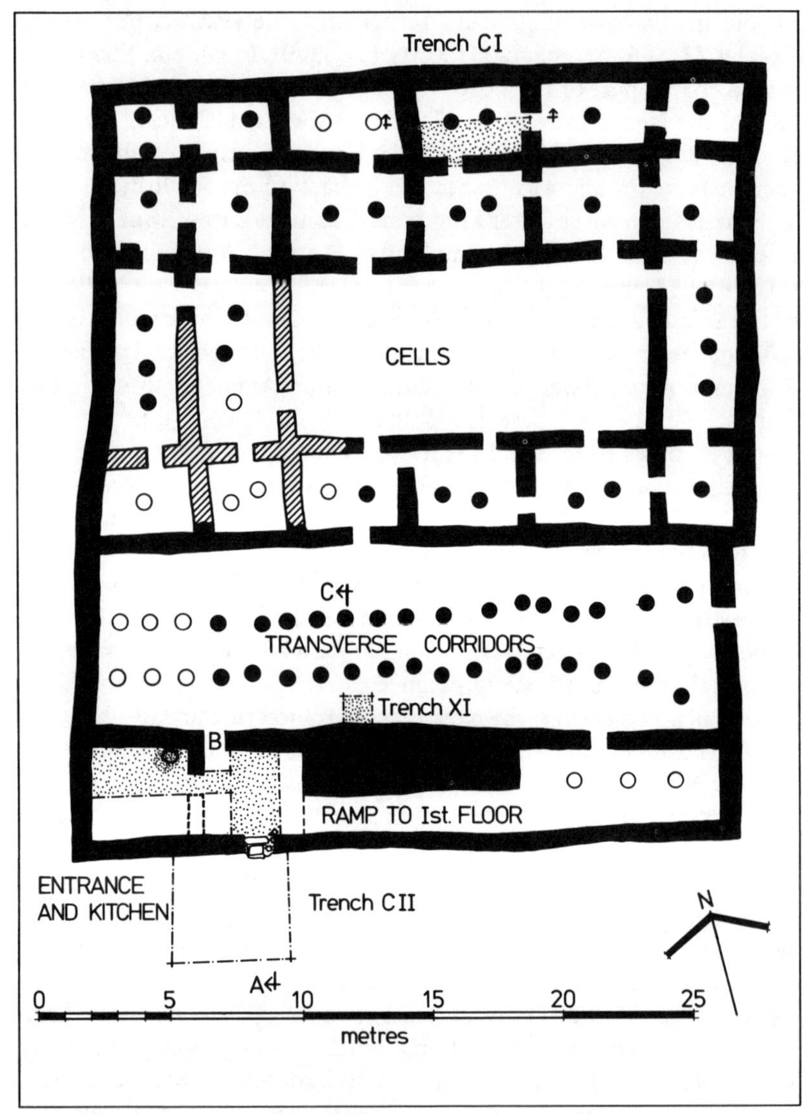

Fig. 14 MBA Citadel: ground floor plan

Presumably the people of that time were able to survive and use parts of the much more ancient water systems that we could still see and record several thousand years later. One might also postulate that the desert was populated by nomads. Thus certain foodstuffs

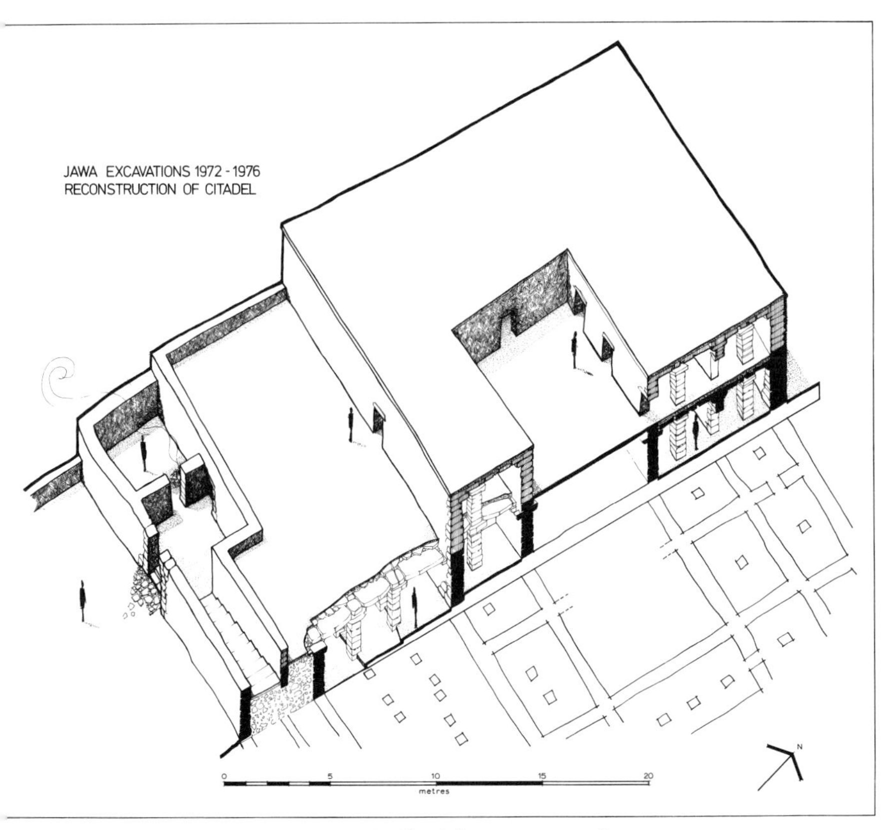

Fig. 15 MBA Citadel: reconstruction

would have been available by barter. As will be seen in discussing the pre-Islamic Safaitic folk and their relation with the Romans in the region, the bedouin themselves had probably been using Jawa as a staging post and a watering place. It may be argued that whoever built the Middle Bronze Age complex was not of indigenous stock but came as an intruder, like the Romans, and from beyond the desert – from urbanized and 'civilized' lands.

In the same way that the bedouin had used Jawa since the beginning of human life in this desert, so, it may be postulated, did the folk responsible for the Citadel and its outbuildings. They would have seen Jawa as a convenient resting place where water was sometimes available in an otherwise dry land. There on the ruins of another much earlier settlement they built a small, definitely not urban complex which must be a Middle Bronze Age caravanserai, perhaps one of many though certainly the first so identified. The complex may have been no more and no less than a hotel on the route between Syria/Mesopotamia and Arabia and Palestine. Since one always looks for more than just stones and subsistence economy, and since a pattern and thus a possibility of historical identity can be recognized, I have called the place Khan Ibrahim.

and Abraham took Sarai his wife, and Lot his brother's son, and all their souls that they had gotten in Haran [northern Syria]; and they went forth to go to the land of Canaan; and [but by what route?] into the land of Canaan they came. (Genesis 12:5)

Whether this is really so or not is less important, as will be seen later, than the potential route implied. So far as our contribution to the history of the Black Desert is concerned, we have at least partly filled the gap between the town of Jawa and the period when the area properly becomes known to the outside world.

And so we have reached the time of the Greeks, the Seleucids and primarily that of the Romans, or *rumi* as they are usually called by the Safaitic bedouin and others. Of these 'others' the Nabateans were the most Hellenized, then Romanized and civilized. Petra was their southern capital. The desert city of Umm el-Jimal on the western edge of the basalt just over 100 km from Jawa, is one of the best known Nabatean and Roman towns in this northern area. But it is only a part of the broader settlement pattern that included the frontier posts (*castri*) of Umm el-Quttein, Deir el-Kahf and Qasr Burqu', to mention just a few, as well as many other towns like Salkhad and Bosra north-west of Jawa.

History proper touches Jawa but slightly in the form of two types of inscriptions found nearby. They derive from the two cultures in the region: the intrusive, urban Roman, and the indigenous, nomadic Safaitic.

For the former the most topical comes from a place called Jathum. It speaks of life in the Bilad esh-Shaytan during the second or third century after Christ, and its content, even its tone, is much like the utterances of many colleagues during the excavations of Jawa:

Life is nothing [or worth nothing]. [As for] Diomedes the lyrist and Abchoros the barber, the two [of them] went out into the desert and were stationed with the commander of the foot soldiers near a place called Abgar. (Mowry 1953)

From Jawa itself we have one of the very rare Safaitic inscriptions that actually refers to historical events of some portent, rather than interesting but in the end uninspiring records of bedouin names:

By Ḥair bin 'Aus bin Ḥair of the tribe of Masikat. He was born in this place [Jawa] the year of the rebellion of Muḥarib and the year of the rebellion of Damasi. He is on the watch for the enemy, so, O Allat and Du-Sara [grant] security and . . . (Winnett 1973)

The Safaitic in the text above, according to Winnett, is dated in the early second century of our era but concerns events–a revolt among the desert tribes – of about the middle of the first century. The name Damasi has been related to a grandson of one Damasippos, a Hellenized Nabatean. Damasi, the story goes, was passed over in favour of his brother Maliku as strategos of Hegra in AD 72. The Nabatean king at that time was Rabbel II (AD 71–106), the last ruler of the kingdom. It is interesting, as an aside, to note that this was also the time of the First Jewish Revolt and the siege of Jerusalem by Titus that is reported so dramatically by Josephus.

There are thousands of Safaitic inscriptions at and about Jawa and clearly that place was an important meeting point on the normal interior desert route which still passes nearby. The content of these messages – which Oxtoby, one of their interpreters, light-heartedly calls the work of a desert Kilroy – shows us the relationship between the nomads and the town folk further to the west. Quite often the implication is that the *rumi* were to be treated with care ('He is on the watch for the enemy') and, conversely, that the bedouin were a mixed blessing, being useful in providing horses and desert produce but also tempted to raid the towns. One inscription reflecting this was found at Jawa during the excavation and reads: 'By MSLK son of BDBL son of SLM and he wrote [or, and he drew]' (see Appendix G). The text (one of three) accompanies one of the finest Safaitic drawings yet discovered in the desert, representing a rider of military bearing. He is shown killing a very nasty beast with his lance, rather like later Byzantine paintings of St George and the Dragon. The beast may well be a lion, for the last, it is told, was killed near el-Azraq only about thirty years ago. But what is more relevant to our desert history is the man and his horse, because they fit the image evoked by the various inscriptions, referring to tribesmen

Fig 16 Carvings at Jawa: mounted man hunting lion

trading horses with towns like Basra. Mention is often made of war as well as trade. The Jawa rider's armament, especially his helmet, might easily have been the fruit of such transactions. On the other hand the man might be a Roman, drawn by an admiring bedouin who coveted his superior equipage.

The terminal date for the Safaitic inscriptions is normally taken to be about the fourth century AD since the latest north Arabian inscription is dated AD 328 and clearly shows a preference for the Nabatean script.

This then is the extent to which the outside world touched on Jawa in the frontier areas between the Romans and their eastern neighbours, the Persians. There were of course many minor political entities, local dynasties whose loyalty fluctuated according to convenience and the relative strength of the greater protagonists. The Persians invaded Syria under Chosroes I Anushirvan (AD 531–79) and destroyed the Byzantine city of Antioch. By AD 613 they were south of Damascus, took Jerusalem and reached Egypt in AD 619.

Attempts were made to regain these provinces, including the marginal desert areas, and to reorganize the frontier fortresses and settlements according to the earlier Roman principles. In AD 627, the Muslim year of the Hegira, Heraclius penetrated to Nineveh on the Tigris river, but Syria and therefore the Black Desert never again saw the *rumi* rule. At the Battle of the Yarmuk on 20 August AD 636 the Byzantine army was decisively defeated and from that time until the eighteenth century no western man who could leave a record for us entered this desert. None came to the Jawa area until the later nineteenth century.

For a relatively short time the old Roman and Byzantine frontier posts and also their towns were occupied and some new castles were built by the Moslem Arab dynasties that followed the end of Roman rule. There are Kufic inscriptions at Qasr Burqu' attesting to this period (Field 1960) and others at many of the *castri* all along the line of the Roman lines. But for the most part the Black Desert of Syria, Jordan and Saudi Arabia remained as always the land of nomads, some of whom were ultimately reorganized – as perhaps in Roman and Byzantine times – as the Arab Legion of Transjordan, to become the bedouin police force in the desert. They were installed in forts that are strikingly similar to Roman *castri*, and at Deir el-Kahf one of their outposts was built only a few hundred metres from the old *castrum* which modern bedouin have made into the core of their new permanent settlement.

The pattern is clear and, it seems, almost timeless. It reflects the ambivalent relationship between the desert and the surrounding verdant lands, the nomad and the peasant (and villager as well as citizen) and without doubt, in an inversion demanded by the environment, the permanence of the one over the transience of the other throughout history – and, as will be seen, prehistory. Only through technology has this pattern apparently been broken in Jordan today – as it was for a while at Jawa over 5000 years ago.

Jawa is therefore the anomaly whose parallel is intrusive urbanism from without. Because of the fate of this fourth-millennium town there is a strong suspicion that Jawa was a freak phenomenon and an accident: a paradox in one very short moment of time.

6
THE 'OLD MEN' OF ARABIA

Recent history has affected not so much the basalt desert itself as ideas about the kind of human occupation that might once have existed there before the Safaitic bedouin: before the Iron Age and before even the Middle Bronze Age. In the preceding chapter it was claimed that there is new evidence about the 4000 or more years between about 8000 BC and the construction of Jawa in the later fourth millennium. In other words, relative to that town we claim to have defined or recognized the existence of an indigenous population in the basalt desert, a people who are possibly the ancestors of the Safaitic bedouin and through them the forefathers of Jordan's present-day nomad population. This may be so certainly in terms of lifestyle, if not in the strictest genetic terms. There is therefore an opportunity to set realistically the human as well as the physical or environmental stage upon which Jawa will appear.

The recent history referred to concerns the decades after the end of World War I when the airmail route between Cairo and Baghdad was established. The new evidence comes from the excavations at Jawa and several related surveys within the basalt region.

This story began with Poidebard's flight over Jawa in 1931. Four years earlier and in the first volume of the British journal *Antiquity* another flyer, Flight Lieutenant R. A. Maitland, published an account of what he and others had seen when flying along the airmail route once the oasis of el-Azraq was left behind. He wrote of strange walls and reported that the bedouin thought them to be very old indeed – even older than the time of the *rumi*. There were small open settlements made up of hut-circles and large walled complexes that seemed to be fortifications of some kind. In some parts there were lines of 'hill forts' linked one to the other by kilometres of low walls; in others, twinned 'forts' surrounded by sprawling enclosure walls

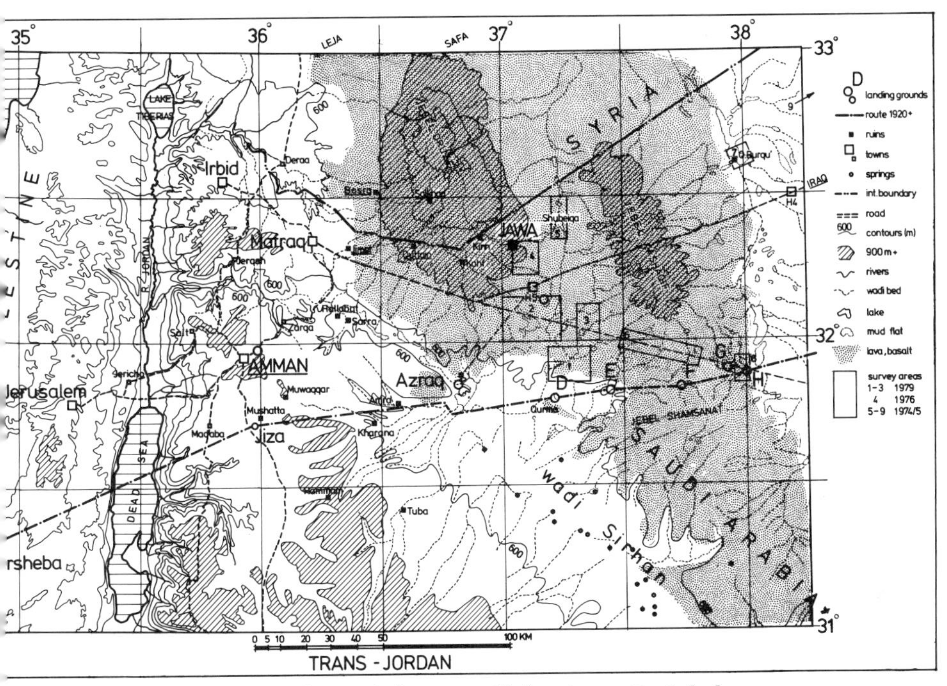

Fig. 17 Map: Transjordan and the 1920s flight plan to Baghdad

which by their extent seemed to imply permanent occupation in the past. One of these, merely identified as a hill fort in Arabia (I have tried to find it, without success), seemed to consist of an upper and lower ring of fortifications rather like Jawa. Maitland compared this place to a Welsh Iron Age hill fort, not without some justification – even though there is a good possibility that much of his 'fort' is a natural volcanic formation.

Maitland also described, for the first time, certain stone patterns on the desert floor that were to become a much discussed enigma potentially rivalling the hitherto more famous lithic wonders such as Stonehenge, Avebury, Carnac and the Nazca mega-portraits of South America, to name but a few whose interpretation has often strayed far beyond the natural. The Black Desert phantoms were star-shaped enclosures from which kilometre-long arms radiated across the landscape. They were grouped in chains by the thousand and implied an almost cosmic purpose. The prosaic pilots who had been reporting these weird shapes called them 'desert kites' and so compelling was their apparent mystery that all other features in this desert took second place in the lengthy explanations that followed over the ensuing forty and more years. Most of these were unsatisfactory, though mercifully none was supernatural. They encompassed the totally wrong, the right for the wrong reason, the limited – through lack of first-hand information – and most recently the tautological. Quite simply, nobody since the 1920s bothered to survey the 'kites' which find their greatest concentration in Jordan's basalt desert. The only exception is a number of meagre examples in southern Israel. As usual, it seems, the Bilad esh-Shaytan was a deterrent and scholars were satisfied with secondary, tertiary and sometimes even vaguer information.

Maitland's account was in many ways one of the best under the circumstances because it was the least biased. He simply reported what he saw from the air without tangential speculation. He said that the bedouin call all of these puzzling features the Works of the 'Old Men' in Arabia: the title of his article and as an acknowledgment that of this chapter.

Let us first take up and answer the long question of the 'kites' and then turn to the other sites and in the end describe them all in terms of a land-use and settlement pattern that pre-dates the advent of Jawa.

There are two basic questions regarding 'kites': their function and

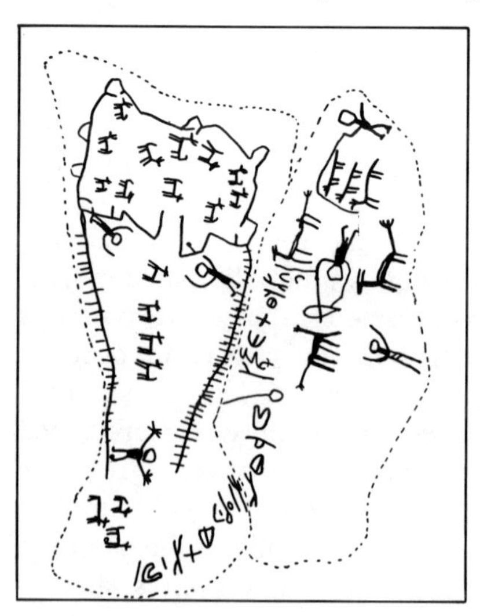

Fig. 18 The cairn of Hani'

their date. A lot of nonsense has been written about both and it is not worth repeating much of it in detail. In terms of function they have been called fortified enclosures with defensive embrasures intended to contain domesticated animals. This view was apparently supported by a chance find which also seemed to provide the answer to the second question of date. Gerald Harding found a carving of a kind of 'kite' accompanied by a Safaitic text reading 'By Mani'at, and he built for Hani'. And he drew the picture of the pen [or enclosure] and the animals pasturing by themselves' (Harding 1953). He was, unfortunately, wrong on both counts.

As to function, two relatively recent ethnographic parallels might be quoted. They should be read – especially the second – in conjunction with the illustrations. The first is attributed to the Amir Nuri Sha'alan in answer to Poidebard's questions about 'kites' and the so called fortified enclosures: 'ces enceintes ne sont pas arabes, mais romaines. Elles ont été employées, même à l'époque romaine, puis a l'époque arabe, comme pièges à gazelles'.

The second parallel was appended as an editor's note (Crawford) to another article about 'kites' by Group Captain Rees, a colleague of Maitland's in the 1920s. In the words of the discoverer of Petra, gazelles

> are seen in considerable numbers all over the Syrian Desert. On the eastern frontiers of Syria are several places allotted for the hunting of gazelles; these places are called *masiade*. An open space in the plain, of about one mile and a half square, is enclosed on three sides by a wall of loose stones, too high for the gazelles to leap over. In different parts of this wall gaps are left, and near each gap a deep ditch is made on the outside. The enclosed space is situated near some rivulet or spring to which in summer the gazelles resort. When the hunting is to begin, many peasants assemble and watch till they see a herd of gazelles advancing from a distance towards the enclosure, into which they drive them; the gazelles, frightened by the shouts of these people and the discharge of fire-arms, endeavour to leap over the wall, but can only effect this at the gaps where they fall into the ditch outside, and are easily taken, sometimes by the hundreds. The chief of the herd always leaps first, the others follow him one by one. The gazelles thus taken are immediately killed, and their flesh sold to the Arabs and neighbouring *Fellahs*. Several villages share in the profits of every *masiade*, or hunting-party.
>
> (Burkhardt 1831)

These parallels, as we were to discover, are relevant to the original function of the 'kites' which was indeed the hunt – not only of gazelles but of antelopes, ostriches and any other game that crossed over the Black Desert – but that is about as far as one can safely take them without actually examining 'kites' in the field. These recent accounts talk about isolated examples, just as Harding's carving depicts merely one 'kite', while we are dealing with thousands, all remarkably similar in design and scale. Near Jawa lies a chain of over sixty 'kites' and we have counted over 100 in chains further to the south and east. The principle was always the same – to drive herds towards a restricted killing ground – but the structural detail was not; and this is where nearly everyone was so sadly mistaken through relying too much on ethnographic parallels and not enough on field work.

Several 'kites' found in 1979 some kilometres south-east of Jawa are illustrated and in these one can easily visualize – aided by

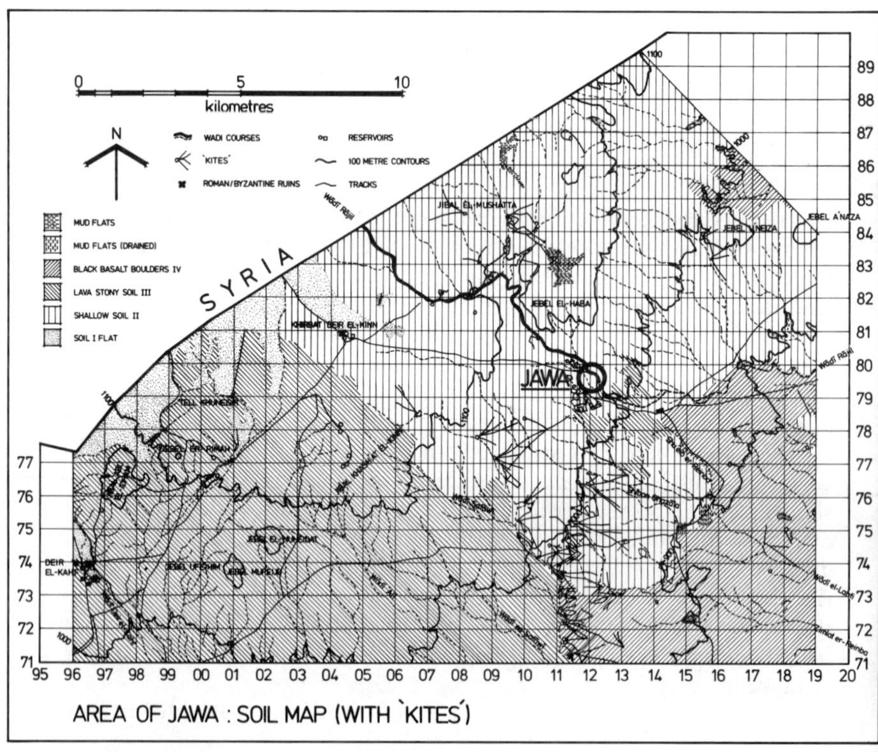

Fig. 19 Map: Jawa area and chain of 'kites'

ethnography if necessary – a herd of gazelles stampeded along the extensive guiding arms that grow imperceptibly closer to each other, always over a low hill or saddle (not a plain as in Burkhardt's hunt), through the neck of the 'kite' into the enclosure that was soon to

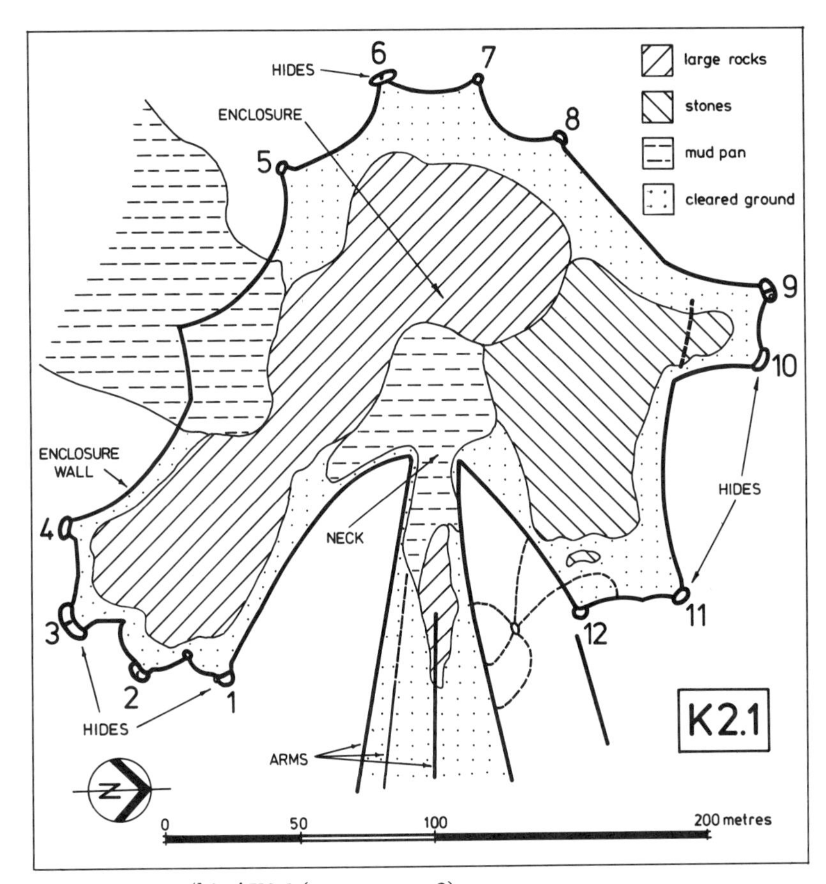

Fig. 20 Map: 'kite' K2.1 (survey area 2)

become the killing ground. The animals would spread out, still carried forward by their frantic momentum, only to meet the farther enclosure wall which was provided with rounded stone hides (figure 20:5–8) – not windows or pits – manned by the archers and spear men, the 'Old Men' of Arabia. Some gazelles would be brought down, others would veer left and right towards the side projections

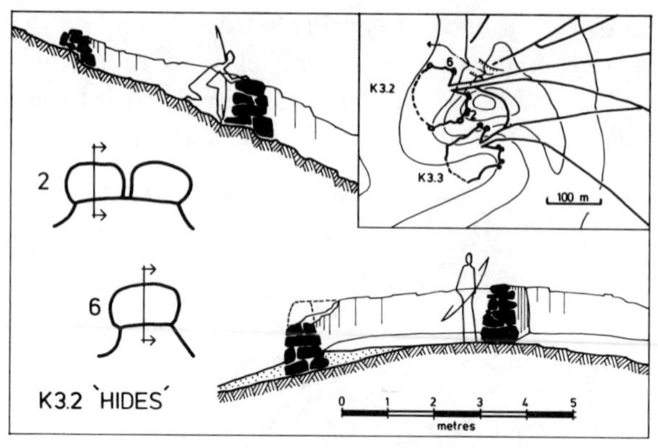

Fig. 21 Map: 'kites' K3.2 and K3.3 (survey area 3)

of the enclosure whose ends also concealed hunters in single or double stone hides (figure 20:1–4, 9–12). The cross-fire would be deadly and no animal would long remain alive. Those that did would be finished off with spears, probably at close range.

The caution expressed here regarding ethnographic parallels should have been extended to the dating evidence of the 'kites'. Harding's stone of Hani' merely points to the use of a 'kite'-like structure about 1500 years ago; moreover the enclosure only resembles the typical star-shaped 'kites' that appear in their thousands in long chains running north-south all the way from the east shore of the basalt desert to el-Azraq in the west. Group Captain Rees and the Amir of the Sha'alan both suggested a Roman date, close to Harding's. More recently another scholar (Yadin 1955) proposed that 'kites' ought to belong to the beginning of the Early Bronze Age – close therefore to Jawa – because he thought he could detect a representation of one on the predynastic Egyptian palette of King Narmer. He was more right, but for quite the wrong reason and his interpretation has been refuted by Ward (1969). What is 'more right' about Yadin's hypothesis is an early date for the 'kites', although this is not proven by the palette, and the existence of an overland route between Mesopotamia and Egypt *via* the Black Desert during the third millennium – something that has always been pretty obvious in any case. No new evidence was therefore provided and, until now, only one archaeologically reasonable proof came from the south of Israel where come 'Chalcolithic' pottery and flints were found in a small 'kite' (Rothenberg 1972). But even this evidence is suspect. In the end this entirely unhappy state of affairs was summarized at length by another Israeli who could of course not examine the best examples and therefore added nothing new (Meshel 1974).

Let us now look at some real evidence. First of all, there is a single small 'kite' at Jawa, downhill from one of the urban water storage areas. It has one 'trap' like the examples from Israel and its guiding arms were altered during the construction of the pool. The 'kite' is therefore earlier than Jawa: earlier than the later fourth millennium BC. Next, south-east of the town lies the long interconnected series of 'kites', each of the multiple-hide type. The last one before wadi Rajil was overbuilt by one of Jawa's animal pens. And thus Jawa itself has provided a *terminus post quem non* to something like sixty

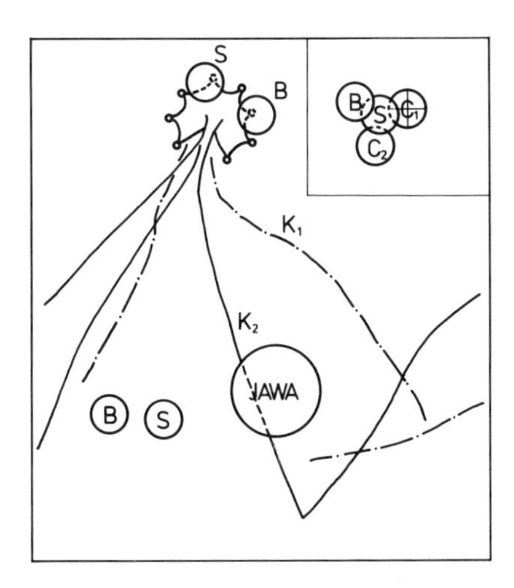

Fig. 22 Horizontal stratigraphy of 'kites'

of these structures; but only if one accepts the (obvious) similarity of the design as a sign of contemporaneity.

To confirm our view we undertook a number of surveys. These at once demonstrated the vexing rarity of artefacts within the 'kites' that could give us a date and dampened somewhat my criticism of others who found no proof, even when they deigned to venture into the Bilad esh-Shaytan. Twice we covered nearly 20 km on foot and found just one flint – and that undiagnostic. True, we found no pottery or any of the other more modern artefacts such as rubber boots or tin cans which are the archaeological proof of modern bedouin occupation; but we had no definite evidence. Then, in 1979, during a survey in the region of Qurma, at Qa'a Mejalla near one of the landing grounds once used by Maitland and Rees, four lance points of the Neolithic period were found in one of the rounded projecting hides of a large and typical 'kite'. At long last real and meaningful data were to hand. This was confirmed several months later in a string of 'kites' some 20 km north towards pumping station

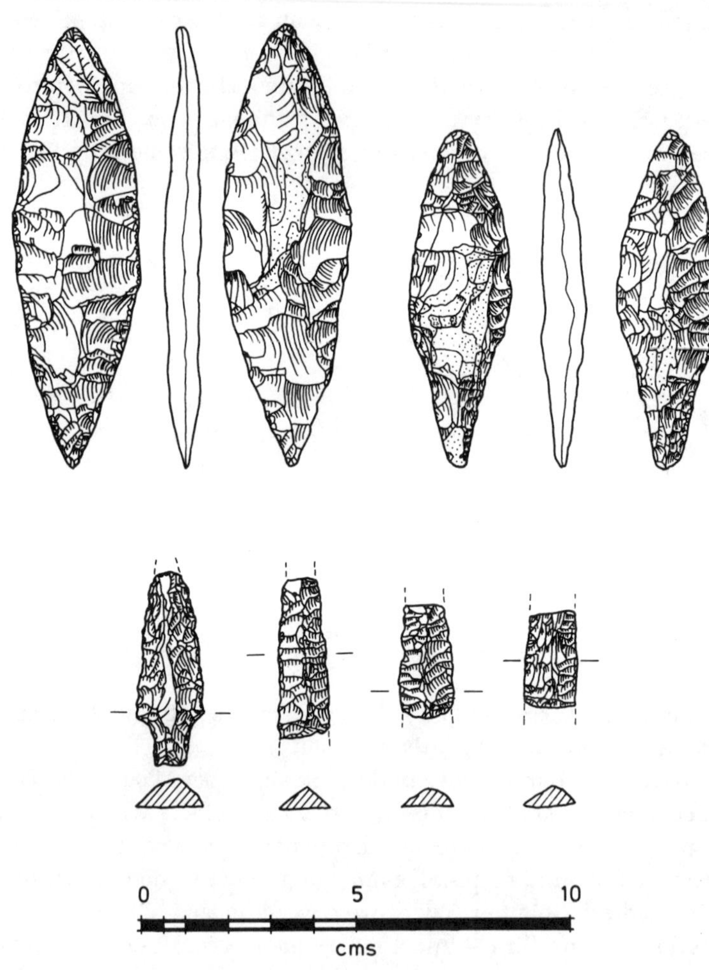

Fig. 23 Flint implements from 'kites'

H5 where diagnostic arrowheads were found in similar contexts (see Appendix C).

As far as the desert 'kites' are concerned, it is now possible to say that they represented a highly organized, co-ordinated and communal hunting technique that exploited the seasonal migration of game across the Black Desert and that they were in use as early as the 7th–5th millennium BC. They obviously continued to be used, though perhaps not to the same degree, and may even have persisted as a means of hunting long after Jawa had been built and abandoned. It is very probable that whoever these 'kite'-builders were – and we may as well call them the 'Old Men' of Arabia – some of their fourth millennium relations would have come into contact with the people of Jawa. The 'kites', once built, would have been ready for the hunt, but for minor alterations and repairs, for millennia, and perhaps only the domestic dwellings of these ancient desert folk would change.

This settles one – perhaps the major – question regarding subsistence architecture between the 7th–5th millennium and Jawa; it is necessary next to consider the other structures that have been more or less ignored since Maitland wrote about them in 1927.

There are two basic forms of domestic 'architecture' that one encounters throughout the Black Desert: rather formalized and regular rounded structures (C1) made up of two concentric stone circles, internal radiating divisions and external hut-circles and (C2) more irregular hut-circles. Of these the second type seems to be the most common and its use continues through millennia of desert prehistory and proto-history, even up to the time of the Safaitic tribes (site type S) and their heirs, the bedouin of recent times (type B). Some of these sites have been surveyed and therefore provide empirical evidence (Appendix C).

In terms of settlement type the rounded sites (C1) may be considered formalized, perhaps extended family units. Type C2 on the other hand can occur in a variety of ways, either singly or in clusters. This leads to a third settlement type (M) that we have not yet been able to reach on the ground; we have only seen these sites in aerial photographs. They are the type that Maitland photographed and called hill forts. These settlement types are summarized in the schematic diagram showing their location with regard to natural features. The hill forts and the regular sites (C1) need no further comment here. Type C2 occurs as follows:

1 single units (one family)
2 multiple units, small camps in the middle of the basalt
3 multiple units against rock scarps for wind shelter
4 multiple units along wadis, often close together
5 multiple units around mudflats, on basalt 'islands'
6 units larger than 2 to 5, larger camps on rocky spurs
7 units like 6, but on isolated hills or on earlier sites
8 apparent expansion of units (4)

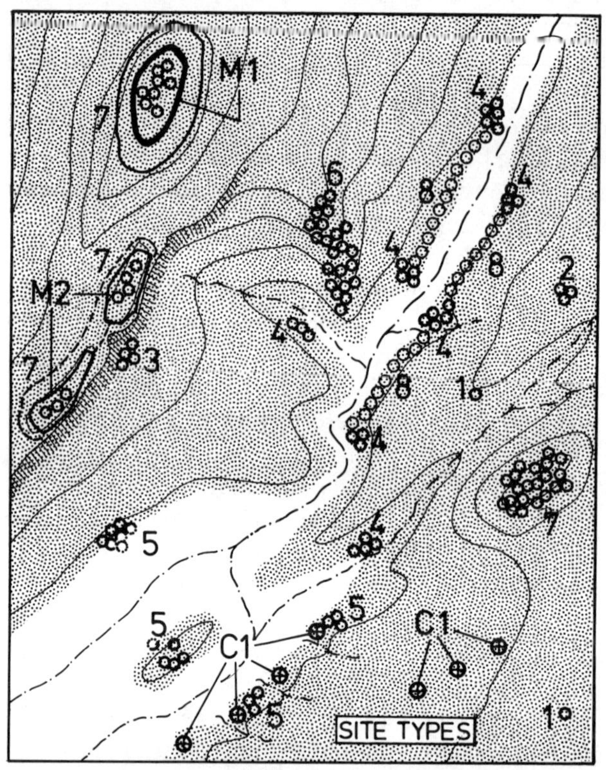

Fig. 24 Site types in the Black Desert before Jawa

It is of course very difficult to determine whether these sites were all lived in at the same time (unlikely), especially types 1–5, or whether (more likely) they were used intermittently. The larger camps are better candidates for permanent settlement, but there is still little proof of this and they would have to be excavated. Both from the air-survey evidence and the multiplicity of unit arrangement, it is clear that the irregular hut-circles (C2) are the most common form of human shelter. They are best understood in terms of the modern bedouin tents, no matter how old they might be. Even their superstructure was very likely fabric or skin rather than more permanent material.

When we began to set up a chronological scale for these ubiquitous sites we were fortunate to find a kind of bedouin Arabian Rosetta stone that illustrates this type of dwelling and underlines the amazing continuity of life-style in the Black Desert. The stone was found off the Trans-Arabian Pipeline road from near pumping station H5 to Saudi Arabia, in wadi Ghadaf. Some of the Safaitic text is cut over a diagram that can only be a cluster of hut-circles such as type C2 (see Appendix G for the texts). Of course one cannot be certain that the drawing is therefore earlier than the Safaitic period (pre-first century BC), but that is unimportant. It represents a typical dwelling that like the modern bedouin sites has a two-roomed living quarter (AA), one for women, the other for men receiving guests, a stone forecourt (B) and several stone corrals (C) for animals.

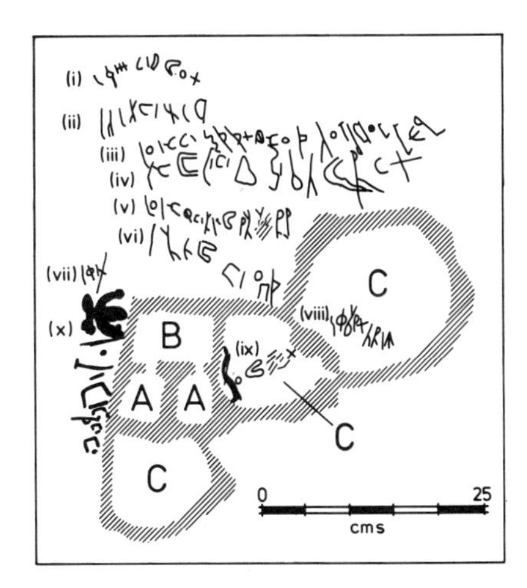

Fig. 25 The wadi Ghadaf stone

A horizontal stratigraphy, quite apart from artifacts, can be established in which bedouin sites (B) are often on Safaitic sites (S) and both either re-use material from the C2 sites or are built directly on them. Bedouin and Safaitic sites are also found built over 'kites' (K) or between the radiating arms, clearly proving that those 'kites' at least must be earlier. Sites of type C1 and C2 are not known to obstruct 'kites' although they do occur nearby. Jawa, as we saw, is later than most 'kites'. No hill forts have been found to obstruct 'kites' (this from aerial photographs alone).

Finally we turn to the preliminary survey results regarding diagnostic artefacts. The finds from site type C1 and many more from type C2 consist of flint tools that date from the Neolithic period: roughly 7th–5th millennia (Appendix C). A similar date for the 'kites' has already been established. There is also a striking similarity

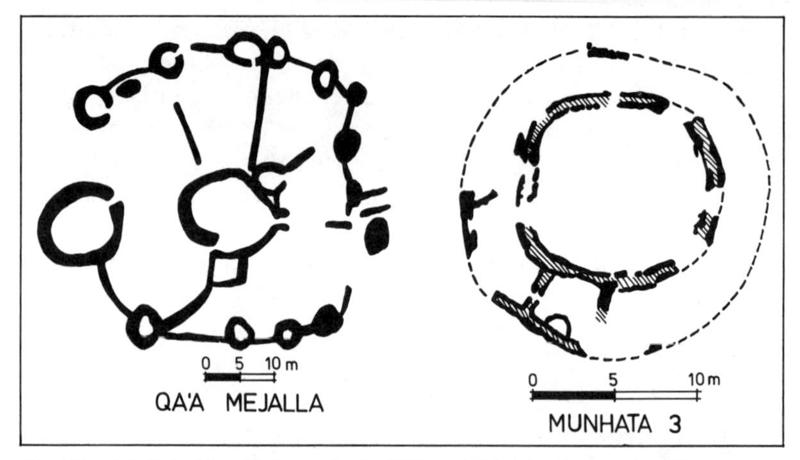

Fig. 26 (a) Qa'a Mejalla hut-circle (C1) and (b) Munhata 3 dwelling

between type C1 (Qa'a Mejalla) and a settlement in northern Israel (Munhata 3: see Perrot 1964) which belongs to the Pre-pottery Neolithic B period of the 7th and 6th millennia. At Azraq I photographed several circular stone sites that might belong to the same category. It is quite possible therefore that our 7th–5th millennia date merely represents a time of use, not the actual beginning in terms of continuous human occupation in this desert. The same was noted concerning the 'kites': that the Neolithic date may not actually be the true beginning of this form of hunting. Since our explorations are still in their infancy, even some of the C2 sites and the (unexplored) hill forts (M) may likewise go back to earliest Neolithic times.

By the late fourth millennium these still essentially Neolithic hunters and herders knew the place Jawa and watered their flocks there after the short winter rains.

III

THE BEGINNING

Sennacherib's siege camp (British Museum): orderly pseudo-urbanism; the city idealized in a camp symbolizing an aggressive civilized imperialism

7

THE ROAD OF THE RISING SUN

In the desert about Jebel Druze there lived a 'large permanent population' whose means of livelihood were hunting and probably also herding. These 'Old Men' of Arabia may have known some forms of agriculture since that science as well as pastoralism was well established in the Near East by that time. In character these people before Jawa might be regarded as fourth-millennium bedouin who roamed the desert much like their modern cousins, year in and year out, exploiting the environment to the limits of their technology and so surviving from one generation to the next. Pre-urban Jawa was at that time a natural rock island in a dominant and hence well-known valley where caves gave shelter to passing nomads, where water collected in natural pools after winter rains and where a time might be spent while grazing the flocks or waiting for the gazelles to pass the line of desert 'kites' to the south. These forefathers of the bedouin may have sat, as their heirs still do, sheltering in the lee of the steep basalt cliffs next to Jawa, and their passing, but for the 'kites', leaves no more sign than the wind. To them Jawa probably meant no more than it did after it had become and been a town, when the semi-literate Safaitic bedouin carved their inscriptions on the stones nearby.

The desert around Jawa is the greater stage upon which the short urban drama was soon to begin. The Bilad esh-Shaytan, dominated less by the brooding mass of Jebel Druze than by the endless wasteland of blackened rocks, yellow white mudflats and chains of extinct volcanoes, supported life for only a few months each year when all the water that could come descended in violent storms that transformed the land. For most of the year man could travel here only if he carried his water with him. His range was limited. If he wanted to stay long in this place he had either to organize supply lines from permanent water sources on the periphery of the basalt or think about storing a supply to last through the long dry season. Unless this was done by importing water the source would have to be the unpredictable and sometimes dangerous floods of winter. This was man's lot and we know his natural resources: grazing for the animals he hunted and for the flocks he kept, a variety of plants and the mudflats in which to grow them, water and the earth's natural form to guide it that had not changed much since the volcanoes cooled. Water channelled along these primeval paths was as fundamental to life as the free air; and water and air seem at times the only elements that make this desert differ from the moon.

Further, just as we are aware of their environment so can we understand their relations with lands that lay beyond the desert across the ageless contact zone between the two basic modes of life, the mobile and the static.

To the west lay the Levantine coast separated from the Arabian Plateau by the great moat of the Rift Valley, the Jordan river and the Dead Sea; to the north, the foothills of the Lebanon and Antilebanon mountains. The west and north were truly verdant: rolling hills with trees, lakes, springs and rivers, lands where cool water flowed all year round; a paradise to desert folk beyond which lay the miracle of the Mediterranean Sea – if they ever saw it. These western lands were settled by an agricultural people which we know as the Chalcolithic peasants whose villages, shrines and workshops have been so closely documented in recent times. Just north of the Dead Sea the site of Tuleilat Ghassul has given its name to a part of this culture and until quite recently one could witness the annual pattern of contact between the settled valley and the deserts to the east. Bedouin would come down from the hills of Moab and Amon, from Edom and the interior of Arabia, with their flocks to graze and water them by the tilled land and permanent settlements. It had been this way a long time and probably as long ago as Jawa. In this manner the folk who built the desert 'kites' would have come because their traps spread west from the basalt, beyond Azraq almost to the suburbs of modern Amman. The same would have been true of the Araba Valley south of the Dead Sea and in general of the whole Rift Valley which is a section of the land route from Asia to Africa.

South and east of the basalt are more deserts. The Nafudh is linked to el-Azraq by wadi Sirhan and its long line of springs that makes this the natural way into the heart of Arabia. The Wudian lead directly to the Euphrates river without any serious natural obstacle other than aridity. In the days of the 'Old Men' of Arabia and before there were semi-permanent settlements here, there would have been some contact between the nomads of the Black Desert and those of the Wudian and through them perhaps a tenuous link with the civilized riverine cultures of Mesopotamia. This is simply a matter of geographical factors determining natural routes. The Wudian tend west-east and border on the north-south tracks along the eastern edge of the basalt which lead from Jauf to the Syrian steppes. Today the paved road from Mafraq makes a nearly straight line for Baghdad.

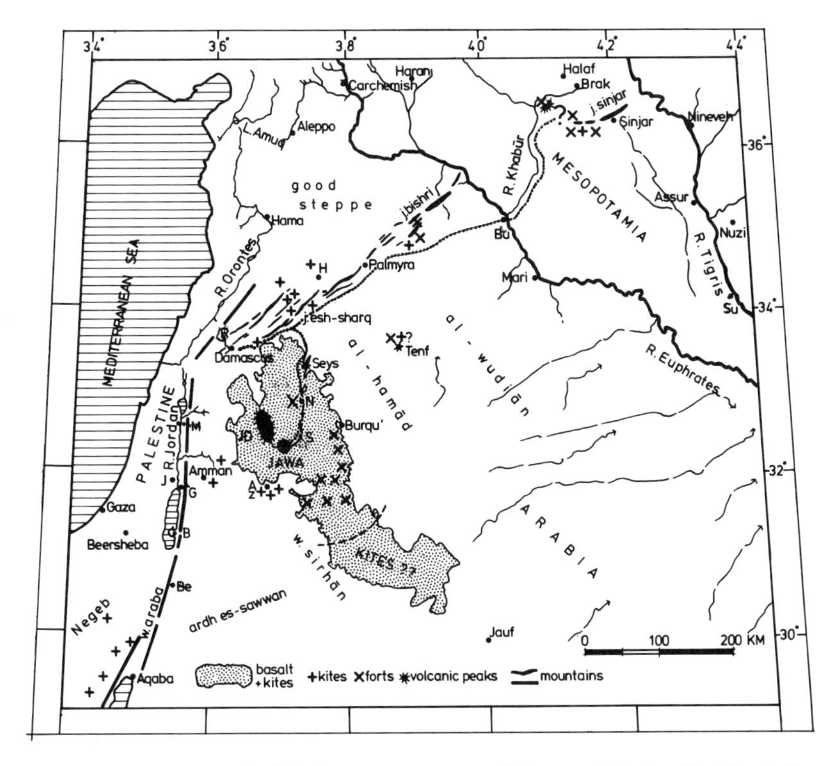

Fig. 27 Map: Road of the Rising Sun. Az el-Azraq, B Bab edh-Dhra', Be Beidha, Bu Bouqras, G Tuleilat Ghassul, H Qasr el-Ḥair el-Gharbi, J Jericho, JD Jebel Druze, M Munhata, N Numeirah, R Tell Ramad, S Khirbet Shubeiqa, Su Tell es-Sawwan

So far as contact during the fourth millennium is concerned, we are certain only of the north and west. At this stage we can assume only an awareness among these pre-urban desert people, a vague knowledge of the more developed cultures of Mesopotamia and Syria. Arabia to the south is still virtually unexplored.

We come then to the north, to western Syria, the Antilebanon mountains, Damascus and the Barada river that made that town an oasis. There we have good steppes and also the nearest urbanized settlements. Archaeological evidence is still sparse and the geographically closest site to Jawa is Hama where levels of the fourth millennium have been found. A very limited narrow sounding has revealed closely packed domestic architecture which might imply urbanism, if only in contrast to the contemporary and obvious villages of Palestine (Fugman 1958). Furthermore, recent discoveries in Syria have considerably altered our impression of the next millennium there. With the hitherto unsuspected literate civilization unearthed at Tell Mardikh (Ebla) we are forced to concede that the roots of the city as a style of organized life are more ancient in this region. Thus it seems possible that the 'Old Men' of Arabia had a direct opportunity of contact with more advanced cultures with all that this implies: trade, diffusion of ideas and warfare.

Most significantly of all, as will emerge, certain geographical features quite apart from the proximity of civilization may have come into play and may have been a cause of change within the basalt desert. These features concern established trade routes that are millennia older than the city.

A long line of hills, occasionally with steep cliffs, runs from Damascus north-east towards the Euphrates river. These hills form the southern border of the good Syrian Steppe and a natural desert route because there are enough water sources along the way. Palmyra is the most famous and part of this region was still called Palmyrena long after the passing of the Romans. To the bedouin this chain of hills is known as the Jibal esh-Sharq, Mountains of the East or even Mountains of the Sunrise. Individually they bear names typical of the bedouin point of view. Those east of Damascus and just north of the basalt (the Safa) are called Rawaq or sometimes Ruwaq. The latter means the flap of a tent used as protection against the wind and can also mean a portico or open gallery. At the northern end of the chain where it touches the Euphrates river the hills are named

Bishri which could come from the words baṣara or baṣira, to rejoice, with the connotation of good omen. If this route is reversed it leads directly from the Euphrates *via* Damascus to the northern gates of Palestine and represents a classical 'invaṣion route' that can be controlled only by holding the water sources along the way and, since these could be circumvented, by making a stand at Damascus where the way narrows between the basalt and lava of the Safa and Jebel Ruwaq.

This route and the one north from Damascus to the Plain of Antioch and the Taurus Range link the extremities of the prehistoric world to the basalt region and along these routes moved the repeated 'invasions' from the earliest times that were to change history more than once. The most ancient facts known about these routes, especially here the eastern one, have come from trade relations during the Neolithic period when a luxury raw material, obsidian, was carried along them from Suphan and Nemrud Dağ beside Lake Van in eastern Anatolia. Obsidian travelled down the Zagros mountains to Chagar Bazar and along the Khabur river of northern Mesopotamia to Bouqras on the Euphrates which lies about 100 km south of Jebel Bishri. From there the road could easily have followed the general direction of the Jibal esh-Sharq to Tell Ramad where the obsidian of eastern Anatolian origin is next found. Tell Ramad is on the Barada river near Damascus and from there it is an easy journey to the Huleh Valley of northern Palestine, past Munhata, through the Jordan Valley down to the Dead Sea and from there to Beidha in southern Jordan.

Both the northern and north-eastern routes are usually reduced to a vague direction 'north and east' when archaeologists are faced with apparently new cultures whose novelty is normally expressed in ceramic or lithic terms. The idea was and probably still is that cultured change is caused by imports which are brought by new folk : variations of the invasion hypothesis. However this ought to be resolved, the current interpretation of material or artefactual evidence is that the late fourth millennium – the time of Jawa – was a period of migrations throughout the Near East.

This pattern and along very similar roads is attested more recently by accounts of entire tribes shifting from one area of Arabia to another and in Turkey we witnessed a similar event whose cause was a catastrophe not within the normal course of transhumance. During the early 1970s water sources had become dry and the villages depending upon them had to be abandoned. The people – easily 1000 – with all their livestock had to move and by the time we saw them they had been on the road for several months. To keep alive they would approach water sources and grazing land owned by more fortunate villagers along their way and negotiate. It was a scene of much historical precedent: of a people dispossessed – by nature this time, not by man – of a people made refugees and nomads who now pitched their tents on the sown lands not their own and asked for terms.

Often they were told to move on after the three days of traditionally free grazing. The flocks would survive by passing from place to place and eating enough at each to stay alive until the next confrontation, which could end in violence.

The cyclical pattern of dispossession, movement and confrontation along established trade routes in the Near East can be traced back through the archaeological record, as was seen, to the time of Jawa which is called either Late Chalcolithic or Early Bronze 1b and by some Proto-Urban. And long after Jawa had passed back to the desert more people were to take to the road for one reason or another and eventually disrupt the cultural, technological status quo of lands far from their origins. A millennium later than Jawa, so the evidence suggests, another such upheaval occurred at the end of the Old Kingdom of Egypt that also marks the end of Palestine's first period of sustained urbanism, both examples of a demographic chain reaction.

The route from northern Mesopotamia and north-eastern Syria to Palestine and Egypt was therefore a much travelled one from the earliest times and thus naturally became a part of various Near-Eastern myths. This was the road that Abraham took – more or less – on his way south from Haran to Canaan, perhaps passing along the Mountains of the Rising Sun. Later a part of this route was trodden by the Israelites of the Exodus – perhaps the most well-known and apposite of stories with regard to the origins and early fortunes of those who built Jawa. That later migration resulted in the far-reaching conquest of Canaan under Joshua and others.

8

WHO WERE THE JAWAITES?

The town of Jawa appeared abruptly. A highly developed and complete life-support system suddenly sprang up in a place where nothing of the kind had existed before. Jawa seems to be the only settlement of its type; and its people disappeared almost as suddenly as they came, after only a short stay, leaving behind them a vast enigmatic ruin. To explain this we have a choice of at least three hypotheses.

The first is that the Jawaites came from a developed urban culture from beyond the basalt region; the second that they invented or evolved a form of urbanism with its requisite technology and that they stem from a people who lived in the basalt like the 'Old Men' of Arabia; and third, that they came from a culture that was developing or evolving towards an urbanism (a proto-urban culture), from a village environment in which certain embryonic elements of the city existed. All three hypotheses would have to include a subsistence technology with knowledge of irrigation and some hydrology.

The nature both of Jawa as an urban system and of Jawa as a rapidly passing urbanism would seem to argue against the last two choices. These ought nonetheless to be explored because they are possible models for change; and they, as well as the first, must explain most (if not all) of Jawa.

Abrupt change has many implications and several explanations. In considering the idea of a city and civilization two mechanisms have been broadly discussed: urban revolution and urban evolution. Both are derived from the saga of the species which is now understood to encompass both mechanisms, forming a dichotomy in the nature of change. Urban evolution is a form of gradualism (*natura non fecit saltum*) and urban revolution an example of what, in modern evolutionary theory, has been called a system of punctuated equilibria. Whether this is applicable meaningfully to the story of urbanization is open to question. There is quite a difference between bio-mechanics and the dynamics of intellect, between cells and ideas. The time scale also is vastly larger; although in proportion perhaps similar to the suddenness of urban Jawa.

We are dealing with changes caused by ideas, the product of human intellect, as much as changes caused by natural growth which can change a village almost imperceptibly into a town. In the realm of the brain revolutionary evolution is quite as possible as conservative gradualism. Thus we know that mental 'quantum leaps' are possible and the causes perhaps parallelled in the physical world. Certain environmental factors can precipitate change, suddenly, just as in evolutionary theory certain pressures will overstress an equilibrium. Something that may be relevant, at least to some of the hypotheses here, is that for the evolution of the species equilibria are suddenly tipped in peripheral areas where isolation favours natural genetic engineering.

Let us now consider the three choices for the origin of Jawa's urbanism in reverse order, starting with the one that seems the least likely: that the Jawaites came from a village environment of the kind flourishing in Palestine, in the Jordan Valley at Tuleilat Ghassul, during the fourth millennium. Gradualism explains reasonably the progress made during the Chalcolithic period from rather scattered, some almost semi-nomadic settlements (even subterranean) to well-ordered rectilinear structures that typify the later stages of the period. All of this took place before the so-called Proto-Urban stage when places like Ghassul were totally and it seems abruptly abandoned some time towards the end of the millennium and their people lost without a trace because their material culture cannot be directly paralleled in the next era. There are similarities, to be sure, and more as archaeologists look for them (abandoning millennial thinking), that suggest some cultural continuity and so perhaps a demographic one although perhaps only because of the nature of containers which never really change completely since their fabric (and form) is virtually constant. In our model here we ought to see the Jawaites' departure from the Jordan Valley near the end of the Chalcolithic period, before the advent of towns, before about 3200 BC or so and certainly before the beginning of the Early Bronze Age. Architecturally we might cite the 'Jawa House' as the generic domestic container unique to the site (figure 59) – unconnected with much more logical structural development for the moment – and so time their departure in terms of Ghassul just before phase IV there because the 'Jawa House' tends to have rounded corners (traditionally primitive) and those of Ghassul IV are square. However one resolves such essentially misleading patterns of change, the story would be that a large group of people and their animals, the former pregnant with impending urbanism (and the right-angle), moved out of the green valley of the Jordan, away from the milk and honey and water, and struck out across the flint-strewn wastes towards an even worse place. Why would they do such a thing?

One specifically technical reason has been cited for the settlement of Ghassul which lies in a region of artesian wells. In terms of ground water resources the site is close to the interface between saline and sweet water just north of the Dead Sea and is thus dependent on a kind of equilibrium. It also lies in a region of frequent tectonic disturbance that can cause wells to shift. Today water is drawn about 1 km from Ghassul which is dry. The ancient settlement could therefore have been abandoned because of a failing water supply.

But why move into a desert that has no apparent water at all?

Assuming that they did leave, and for whatever reasons moved east, the first proper permanent water that they would have encountered would be either the wadi Sirhan wells or the oasis of Azraq. At the latter they would have stopped because, logically and very simply, if they were capable of Jawa's technology (at least potentially so), and in particular military aspects, they would have been more than a match for any bedouin consortium that might have contested their rights to water. Jawa is quite far away into the Black Desert and there seems no compelling reason why these people should have gone there. Also, of course, they did not stay at Azraq long: there are no signs whatever of any early proto-urban settlement. But assuming that they did go north, against all good advice, and that they reached Jawa, why is their material culture only passingly reminiscent of the Jordan Valley they had so recently left?

More relevant than these questions is the one concerning why they immediately created an urban technology so totally different in scale and concept from their recent village roots. Can this perhaps be explained in terms of revolutionary evolution? Should it be interpreted as a tipped equilibrium and the causal stresses the harsh physical environment of Jawa? Perhaps – but not probably. This idea will reappear at the end of the Jawa story when we come to the abandonment of the town and see that the model under discussion now would involve a paradox of improbable intricacy, not to say ultimate irony: a failed urbanism replanted in the land whence it came – a return of the prodigal village, somewhat changed through hard experience.

The second choice again raises the question of gradual or revolutionary change. Did Jawa grow out of the desert? Was its source the Black Desert, or the sands of Arabia, or even God's own Wilderness, Sinai? In all of these areas, with varying degrees of concentration, are to be found the 'kites' of the 'Old Men'. Could they be the Jawaites?

Since the questions are more or less the same whichever desert is chosen, we can limit ourselves to the basalt. Among the others, Sinai has produced at least one quite large settlement of similar date to Jawa (figure 56); Arabia has not yet produced anything comparable. In the basalt, then, is there any evidence for either or both paths towards Jawa's urbanism? The answer is similar to the one in the discussion above: that we can see a combination of both in this hypothetical situation. It will be seen that whatever pattern one might build within the basalt, that pattern is not confined to that place, necessarily forcing an adjustment.

A summary was given in Section 2 of settlement types so far discovered in our area and in them, if the hut-circles (C2) can be linked with the hill forts (M), a case may be made for evolution towards larger sites and some forms of simple fortification. The hill forts – if they really exist – are vaguely similar to Jawa in that they appear to have two lines of enclosure walls. The step from one to the other is explicable in gradualist evolutionary terms. But the leap from these hill forts to Jawa is quite another matter. It could of course be reasoned that the 'kite' folk knew their desert and its resources and had developed water-storage systems. One must however question whether the technological gap between this and Jawa can be bridged. Nothing of the scale or scope of Jawa's water systems has been found elsewhere in this desert. Is a revolutionary explanation adequate? Perhaps the reasons were economic, notably trade with extra-basalt urban societies, one sign of which might be the sudden appearance of pottery at Jawa. It could be argued that trade would produce surplus wealth which made a town like Jawa feasible and that the new prosperous centre of the new economic system was sited at the core of the region, suitably beyond the range of its urban neighbours. Jawa, the mercantile capital of the Black Desert, was safe from the military point of view, but precarious in terms of its own life support and thus a stressed system. Is this the stress that caused the leap to a new plateau of developed urban technology?

This is difficult to believe. The image of water at Azraq and Qasr Burqu' haunts such hypotheses. Why Jawa? Who in the fourth millennium in Arabia and Jordan would or could pose a threat sufficiently serious to force a strong, wealthy, talented and technically advanced community to cling to the edge of subsistence? What fools (the Jawaites) to invest so much in something so short-lived.

Only one thought need be retained once this hypothesis is abandoned as equally unlikely as the first and that is this. Whoever built Jawa did not discover Azraq or Burqu' until after the town was built.

Since no desert is ever completely sealed off and since we are dealing with prehistoric consciousness, with ideas that are at least as portable as pottery or obsidian, we inevitably come to the third choice and the possibility of an extra-desertic, post-village, urban origin of the Jawaites. This is, after all, the easiest way to explain the precipitous appearance of Jawa as well as the startlingly high level of technology. All of Jawa's attributes, the ones that we have been trying above to coerce into believable models, are satisfied by the obvious reality of precedent. One can be sure that developed urban technology existed in Syria and Mesopotamia and probably also in Egypt well before the advent of Jawa in the Black Desert. The new town, moreover, was occupied by people who obviously did not hesitate long – they could not afford to do so in any case – who did not think about choices or what to do next, who did not sit there 'developing' along slow and predictable lines and wait to be bombarded by equilibrium-tipping environmental stresses, but by a highly organized and motivated people who instinctively, as if by conditioned reflex like beavers or ants, began at once to build according to a grand plan. And that plan was part of an idea, even an ideal perhaps, that they had brought with them into the Black Desert.

But, if the Jawaites were so smart, why did they choose such a horrible place? Were they compulsive masochists? How did they manage to miss the water at Qasr Burqu' on the east edge of the basalt and why did they not go to Azraq farther south and west? Why would they not have gone on to liberate the Promised Land from the failing Ghassulian societies who could hardly have resisted a technological *Drang nach Westen*?

One has to ask even before this why they left their urban homeland in the first place; and since they must have gone in large numbers, as one group, how they were able to get away with all of their possessions as well as their lives. Presumably a misunderstanding of some kind caused them to be expelled and to be sent forth unharmed because whoever had the power over them was compassionate but unrelenting, like the jealous God of the Old Testament, and 'placed at the east end of the Garden of Eden Cherubim, and a flaming sword which turned every way, to keep the way of the tree of life' (Genesis 3:24). Or perhaps one should try to be more specific and evoke the frayed yet true models of drought, famine and war, the Apocalyptic vision of the pale horse:

> and his name that sat on him was Death, and Hell followed him. And power was given unto them over the fourth part of the earth, to kill with sword, and with hunger, and with death, and with the beasts of the earth. (Revelations 6:8)

> the kings of the earth, and the great men, and the rich men, and the chief captains, and the mighty men, and every bondman, and every freeman, hid themselves in the dens and in the rocks of the mountains. (Revelations 6:15)

The mountains were those of Bashan, Jebel Druze, at the centre of the Black Desert.

Drought is a definite possibility as was seen in the example of forced migration in Turkey earlier in the story. But why choose a desert in which to rebuild urban life lost? In favour of plague it could be argued that a pestilent city full of blackened corpses would be readily exchanged for a clean though dry desert of black stones. All we are doing here is running through the list of normal causes for migrations and of them the most common is war. It is the recurring pathetic pattern of human failure and human displacement and of the refugee in history who is significant only when the experience does not crush him to silence. He is a sorry footnote otherwise, and it could be argued that the Jawaites were significant because they have left a record, that they were successful failures who could emigrate alive. But the question of why remains.

Both questions – why leave, and why choose Jawa? – have their monumental 'postcedent' in the history and literature of the Near East which is best represented in the pyrrhic victory of Egypt over the Israelites and the story of Moses and the Exodus. That is probably the best known and in the long run the most successful failure in the world and in that story may be seen the Jawaites in whose very existence can be found the roots of such sagas, in the grand demographic pattern of the Near East.

> Now there arose a new king over Egypt which knew not Joseph. And he said unto his people, Behold, this people of the Children of Israel are more and mightier than we: Come on, let us deal wisely with them; lest they multiply, and it come to pass, that where there falleth out any war, they join also unto our enemies, and fight against us, and so get them out of the land.
>
> (Exodus 1:8–10)

Here were a change of government, loss of accepted contacts, virtual assimilation and comfort, wealth and special privileges given a select people who were nevertheless apart and separate from the nation. They were thus a threat, real or not, expressed in population figures (real or not) and an appeal to nationalism. Prejudice and its justification led to an easy persuasion, 'dealing wisely' and all that such words can and have meant. And because there were so many of them the wisest course was expulsion rather than extermination. The historical parallels are as obvious as they are real and (wisely) Moses never lost touch with the desert and kept flocks there and brought to his people their God's great promise of a land for themselves which was at that time (obviously) in the possession of others – the Canaanites, and Hittites, and Amorites, and Perizites, and Hivites, and Jebusites and ... Where would they go when the promise was fulfilled?

Thus the dream, the idea, was consolidated and whether it preceded the threat, whether it was the cause, or whether it became the explanation afterwards, no longer mattered. Less through plagues and floods and famines than the reality of prosaic, economic facts and sheer numbers it seemed best to let these people go. Accordingly they were told:

> Take your flocks and your herds . . . and the people took their dough before it was levened, their kneading troughs being bound up in their clothes upon their shoulders . . . and they borrowed of the Egyptians jewels of silver, and jewels of gold, and raiment . . . and a mixed multitude went up also with them, and flocks and herds, even very much cattle [into Sinai, the desert, across the Red Sea] . . . and the Lord went before them by day in a pillar of cloud, to lead them the way [and after some bitter wells] . . . they came to Elim, where there were twelve wells, and three-score and ten palm trees: and they encamped by the waters.

They were still in Sinai and remained there for decades because all the roads to the Promised Land were blocked by kingdoms and cities unsympathetic even to their just passing.

So similar to the hypothesis of the Jawaites' fortunes is the story of the Israelites that but for the unfortunate illiteracy of the former one could be speaking about the same people. One could go on quoting from the Bible, changing the names and but slightly the places passed. Even before the Exodus Joseph knew the desert routes which Moses and the congregation then trod on their way 'home', just as the Jawaites might also have struck out along ancient and established trade routes when they were exiled from the land in which they were no longer accepted guests.

Let us return therefore to the great north-eastern road spoken of earlier, the Road of the Rising Sun from Damascus past Palmyra to the Euphrates river, and equate Egypt with Mesopotamia, the Nile with the Euphrates, Sinai with the Wudian and el-Hamada, and Elim with one of the oases, like Palmyra. Let us imagine what may have happened once the Jawaites faced their obstacles along a similar path.

To the king of Edom Moses sent messengers, saying:

> Let us pass, I pray thee, through thy country: we will not pass through the fields, or through the vineyards, neither will we drink of the water of the wells: we will go by the king's highway, we will not turn to the right hand nor to the left, until we have passed thy borders [and so reach the Promised Land].
>
> (Numbers 20:17)

But neither for the Israelites nor for the Jawaites was this yet to be.

The population of the basalt desert prior to Jawa has already been described, as have the 'Old Men' of Arabia and their desert 'kites' as well as their camps and perhaps small villages. As with most peoples, their roots probably lay outside the basalt and perhaps their earlier exodus or diaspora or whatever one wishes to call such movements may have followed similar routes. Their distinctive signature on the desert floor, the 'kites' and so-called hill forts, may now give us a clue as to their origins as well perhaps as that of the Jawaites.

'Kites' and hill forts are not found in the Black Desert alone. Poidebard discovered this a long time ago, although none seemed to take note. The signs of the 'Old Men' of Arabia can be traced along the Road of the Rising Sun, from Damascus to Jebel Rawaq and even further north. Past Palmyra and Rujm es-Sabun (Suhne) a 'kite' was found along with many hill forts all on the edge of a cliff just like those first reported by Maitland and Rees in the 1920s. There are more 'kites' in the Suhne area and forts with rounded 'towers'. This takes us past Jebel Bishri and the River Euphrates and south to the confluence with the Khabur and the road into northern Mesopotamia. In the western foothills of the mountains called Jebel Sinjar the same 'kites' and forts were recorded by Poidebard. Suitably, there are also volcanic peaks and lava flows nearby to the north and west of Tell Brak.

Reverse the direction and consider the Jawaites up to Damascus and a long line of people, their flocks and herds, their 'kneading troughs ... bound up in their clothes upon their shoulders' and much else besides – 'even cattle' – stolidly moving from one well to the next along the southern edge of the good Syrian Steppe. They would pass through the desert with the Wudian on their left hand and in their minds they would carry the idea of a new land with water and green hills and fields and grazing unlimited. All along the route would be negotiations for water. They would come to dried-up wells and others blocked or poisoned by people who feared their passing less than their sojourn. And after conflicts and raids at last the congregation would reach the foothills of Jebel Ruwaq, with the black mass of Jebel Druze and the wasteland of the Black Desert on the horizon to the left. Before them lay Damascus and the waters of the Barada river; and the road to this 'paradise' would probably be blocked.

Then they might have sent messengers as Moses was to do who

would walk across the dry dusty soil towards the camp of the people of 'Damascus' saying: 'let us pass, I pray thee, through thy country'. And Damascus would answer: 'Thou shalt not pass by me, lest I come out against you with the sword' (Numbers 20:18). The Jawaites may then have said: 'We will go by the high way: and if I and my cattle drink of thy water, then I will pay for it: I will only, without doing anything else, go through on my feet' (Numbers 20:19). To which the answer came: 'Thou shalt not go through', and the Damascans 'came out against [them] with much people, and with a strong hand' and the Jawaites were turned away.

Back along the road along which they had come, back into the Rising Sun? In the biblical story (Numbers 20:22) the Israelites next journey to the Mountain of Hor, and so in our parallel tale, having been rebuffed before Damascus, the Jawaites may have come to an extinct but impressive volcano, Jebel Seys, which was black and steep and a place where water from winter rains collected in wide pools. From this point on lay the Hamada, the Wudian, the route they had passed over, and their choices were few as to which way to turn.

Quite possibly, like so many people after them in these lands, the Jawaites now crossed over the threshold into yet another and yet harder desert, into the Bilad esh-Shaytan, the Land of the Devil, under the shadow of the black mountain of Bashan. Their road would have been painful as they moved south, but not impassable, especially if there had been some rain. From Jebel Seys runs a long series of linked mudflats. They are fed by winter rains on Jebel Druze and represent a chain of fertile grazing land, even into the dry season, that ended in the south at a place called Khirbet Shubeiqa. On the right would be Jebel Aneza and the foothills of Jebel Druze that were getting lower as the people moved on; but they would offer no solace and still lie as black and alien as before. Below Aneza lay the rock island of Jawa, beside the largest valley within kilometres. Wadi Rajil may have been in flood if the Jawaites came in winter, and since it flows into the Shubeiqa mudflats this lifeline could have seemed like a miracle: where water gushed as if by magic from the dead black stones. 'Behold, I will stand before thee there upon the rock', the leader of the people may have said, 'and there shall come water out of it, that the people may drink' (Exodus 17:6). Jawa had been found.

9

THE FIRST DAYS

Whichever road they took – and there are very few alternatives – they had come in large numbers. Wherever we excavated down to this first phase of Jawa we would encounter thick layers of occupational ash, and where this was found near the fortifications it always ran beneath them. However, we have established that a very short time elapsed between this phase and the next, when the town and its water systems were finished. This allows us to enter even more into the prehistoric urban drama.

The Jawaites had come with their flocks of sheep and goats and their cattle. All needed water now that they had stopped at Jawa, the place that was to become their permanent home. They needed water as they had done all along their hard road; and that one word, water, would have been the most often heard among them in those first days of Jawa.

Because it is dry here for most of the year certain limits can be set on the time of arrival of the new settlers. They would not have come in the summer months of May to September because there would have been no reason to stop except briefly from exhaustion. A summer arrival in wadi Rajil would have forced them onwards and if they had come from the north, as we suspect, they would surely have gone on to find Azraq which is just 55 km away: about two to three days' march. Once there they might even have continued westwards to the Jordan Valley and Palestine. But, obviously, they stopped at Jawa, and these dispossessed folk have presented us with a paradoxical situation in which an undoubtedly intelligent and technically advanced people stopped short of a relative paradise to toil in the hard shallow soil of the Black Desert and to achieve a great but in the end needless technical brilliance during the latter part of the fourth millennium. Conversely, their effect, however small, on

Palestine – as yet not urbanized – was thus delayed by an accident of space and time.

They came to Jawa some time after October and found water. It was either flowing in wadi Rajil, guiding them westwards from Shubeiqa, or (more likely) lying in natural pools next to the site. The closer to the previous summer one sees them arrive the harder their ordeal would have been along the road to Jawa. Lack of water and severe rationing along the route would have made an end to their exodus the more imperative. It is very tempting to imagine those folk turning west for the second time since Damascus and following the living water of wadi Rajil up towards its source in the black hills. And even if the wadi was dry at that time local bedouin might have told them of pools at Jawa. So, as they moved up the valley the wadi bed became increasingly deep and cut into the basalt layers about it. Shortly after passing along two such gorges they would reach a bare rocky mound, an island in the valley as the name Jawa actually implies through its pre-Islamic connotation of 'centre of the valley' (*jiwa'*). The rock had in it many caves offering shelter and we have found the signs of ancient occupation within them. On the west side of Jawa they found a smaller subsidiary valley with two shallow pools that were fed by the rain falling on the land nearby. These pools may have held some water by the end of October and certainly would do so thereafter. One can imagine the scene of milling humanity and livestock as the rigid discipline of the long march broke at the sight of water and the thought of arrival at the end of the road, no matter how harsh and uninviting the place.

In the mud and chill of early winter, strong bitter winds so typical of this land tearing across the flat open desert unopposed by any obstacle all the way from Arabia, they stopped and made their first homes at Jawa. The ashes are scattered thickly everywhere, the remainder of the first campfires on that day as the Jawaites piled high the brush and scrub they found growing on the site and set it alight to keep warm. Among the myriad of smoky flames simple huts and temporary wind-breaks would rise and caves would be swept clean in preparation for the first night at Jawa.

The Jawaites faced all the traditional problems of new settlers in an alien land and the biggest would have been organization and control of the people, the congregation. Discipline, order and strong leadership must all have existed by virtue of the long march to Jawa and because the Jawaites survived to build what is to date the earliest (complete) life-support system ever to be discovered in any desert of the world. One is reminded of biblical parallels: a concentration of mythical, cultic, genetic, tribal loyalty and cohesion, desperation and a strong will to survive.

Their first task on the second day of Jawa would have been to provide shelter. The caves were not sufficient. Presumably they had tents or some form of portable shelter that they had used throughout their travels. But Jawa is cold in winter and even in the summer the temperature can drop to near freezing point at night despite the tropically oppressive days. The wind at Jawa is incessant, especially in winter. So it is likely that a start was made on the rock island that was to become a town so soon and that the small rounded huts, already like those of the developed stages at the site, began to appear among the smoking ashes of brush fires and scattered tents. Most building materials were there on the spot: stones everywhere and mud from the pools for bricks, mortar and plaster. Even some reeds may have grown nearby for temporary roofing until more solid materials could be obtained.

During these first days scouting parties would have pushed on northwards along the wadi Rajil, perhaps to find its source, and after a day or so would have discovered the famous oaks of Bashan, some of which were used in the houses of Jawa. On their way contacts with whatever indigenous people there may have been would have been inevitable; as inevitable and natural as the realization among the Jawaites of the nature of this life-giving valley along which they walked. They would have understood that no springs existed anywhere and that all the water there was came from the surface, after rain.

Food was as essential as the shelters now rising everywhere. We have supposed that they had livestock and that like all such migrants they lived off their mobile living larder, taking care not to slaughter too recklessly. Now whatever remained would become the core of new flocks, the breeding stock, and something had to be done to ensure it. There would have been grazing around Jawa at that time of the year and rough stone pens appeared more or less as we still find them west of the site today. This source of food had now to be very carefully rationed during this initial, fragile stage of the settlement. Similarly, the question of cereals and seed crops must have arisen.

Some supplies might still have been left after the march, especially if they had been replenished by harvesting the ephemeral fields of the indigenous population. If they still had a surplus, this would now go into the thin but fertile soil on the rolling hills west of the site in expectation of an early spring harvest. But it is not difficult to imagine that supplies were low and that therefore peaceful contacts with the aborigines of the desert was desired, quite apart from any other considerations that usually lead to a temporary peaceful coexistence.

In their wanderings the Jawites had probably always been in touch with the local population of the land and meetings would not necessarily have been friendly. From the time that they turned south at Jebel Seys they must have been aware of occupation sites, rings of stones (as we see them on our surveys), recent animal droppings, even footprints in the desert dust. They might have seen small bands of nomads. And if the Jawaites passed through these lands in the late summer, besides replenishing their seed stocks from bedouin sources, they would certainly have negotiated for water, peacefully or otherwise.

From within the desert, through the eyes of the 'Old Men' of Arabia, the picture would have looked equally as mysterious at first, followed by the realization that a large and thus dangerous group of alien people was entering their land. The fact that the 'invasion' succeeded reflects not only upon the relative strength of the potential protagonists, but also on their willingness, if not need, to co-operate.

Qualifying this rather naive view, one must understand that the advantage probably lay with the Jawaites simply because they were organized and the 'Old Men' were of neccessity scattered over their land since their livelihood demanded it. They might have been capable of unification and co-ordinated reaction, but incapable of doing this quickly and therefore in sufficient strength to deal with the Jawaites in any 'wise' way. Taking this further, it is also quite possible that a unified attack would not have become a sustained war because nomads are usually reluctant to engage in the prolonged concentration and discipline that such an action would require. They have always preferred the quick raid and instant tangible profit. Thus isolated attacks might have occurred, not for any vague chauvinistic reasons but for immediate gain, triggered by a traditional and natural suspicion of intruders. However, it is more likely that the first contacts were peaceful because the natives would come in small groups. Curiosity would have been as much a motive as anything else and would have drawn them to these new folk at Jawa. One might argue, therefore – regardless of what may have happened later on – that the initial intentions of the 'Old Men' might have been no more than that and that if they had any other, less social ideas they came to nothing. By the time any indigenous reaction could form into an effective fighting force Jawa, like many other invasions, had become an established fact in the Black Desert.

The nature of contact, the classic primeval historic scene of human confrontation across cultural and sub-racial differences, is perhaps the most interesting aspect of the Jawa story. Unfortunately it has left no literary record and that is why I buttress this prehistoric saga with the Bible and choose to see in Jawa and the Jawaites one of many primal biblical sources. Much has to rest on interpretation and a certain amount of speculation. However, we may proceed with confidence because unlike any other archaeological site that deals with the important question of urbanism near its roots Jawa can be reconstructed at least in physical detail because it is so well preserved. Perhaps it is a dumb record in the literary sense; it is nevertheless an eloquent one.

Yet the question of why Jawa was picked as a site for permanent settlement remains and will continue to do so throughout the story as the second dominant question, of how the Jawaites were able to survive, is explained. This is tied closely to the environment in which they had to live and the exploitation of the meagre natural resources available to them. Of these water was the most vital.

How could they have managed this monumental feat in such a place and in such a short time? And we might ask another question: just how limited in time is the probable moment of their arrival at Jawa?

If they had come in January or February – the most probable time of flooding in wadi Rajil – flowing water might have served as a practical demonstration of the region's resources. But what in this empirical evidence could have informed them, or who could have warned them, of the ephemeral nature of this abundance? Perhaps they had the technology because they came from an urban background, perhaps even a similar area, and therefore understood the hydrology of Jawa. Possibly they learned as they wandered in this marginal world on their way to Jawa and saw how water flowed overland after and even during the rains and that this was all that there would be in any one season. On the other hand such environmental intelligence may have come to them through the inevitable contact with the native population, for then as now the major topic of conversation, especially at awkward first meetings, was the weather. The bedouin could have given information about wadi Rajil: that water flowed for so many days and then no more. Either way Jawa was chosen, and as we were to learn during the excavations this site is the only one of the very few defensible places in the entire region where such water resources could be harnessed successfully in order to sustain a large population throughout the whole year. To a large extent this implies the hypothesis of an urban origin for the Jawaites whose sudden achievement gives the impression of an imported technology. There was an almost instant recognition of topographical potentiality, resources and methods of exploiting them.

But could a large population have survived if the estimated time of arrival at Jawa was during the middle of the flood season? Could adequate water systems be built in time to exploit wadi Rajil? And what would happen if the Jawaites arrived after this – say March or April? Could they have held on and adjusted to a much reduced water consumption rate for up to seven or even eight months, and that after the ordeal of exodus?

However we may resolve this, the fact remains that they had to use what they found at Jawa, at least during the first month or so; and that was the Jawa of the 'Old Men' of Arabia.

Jawa was known to the 'kite' builders or whoever their descendants were just as the site was still a focal point during the Safaitic bedouin period. The topography in the area west of the future town, the area that was soon to become the municipal water storage, can be reconstructed with some accuracy. There must have been two shallow pools collecting surface water from the land immediately to the west and north. Here the 'Old Men' would have watered their flocks during the wet season. Let us therefore reconstruct Jawa's pre-urban water storage potential and set this against a hypothetical consumption rate of 500 m^3 of water per month which reflects a population (with flocks) of about 2000

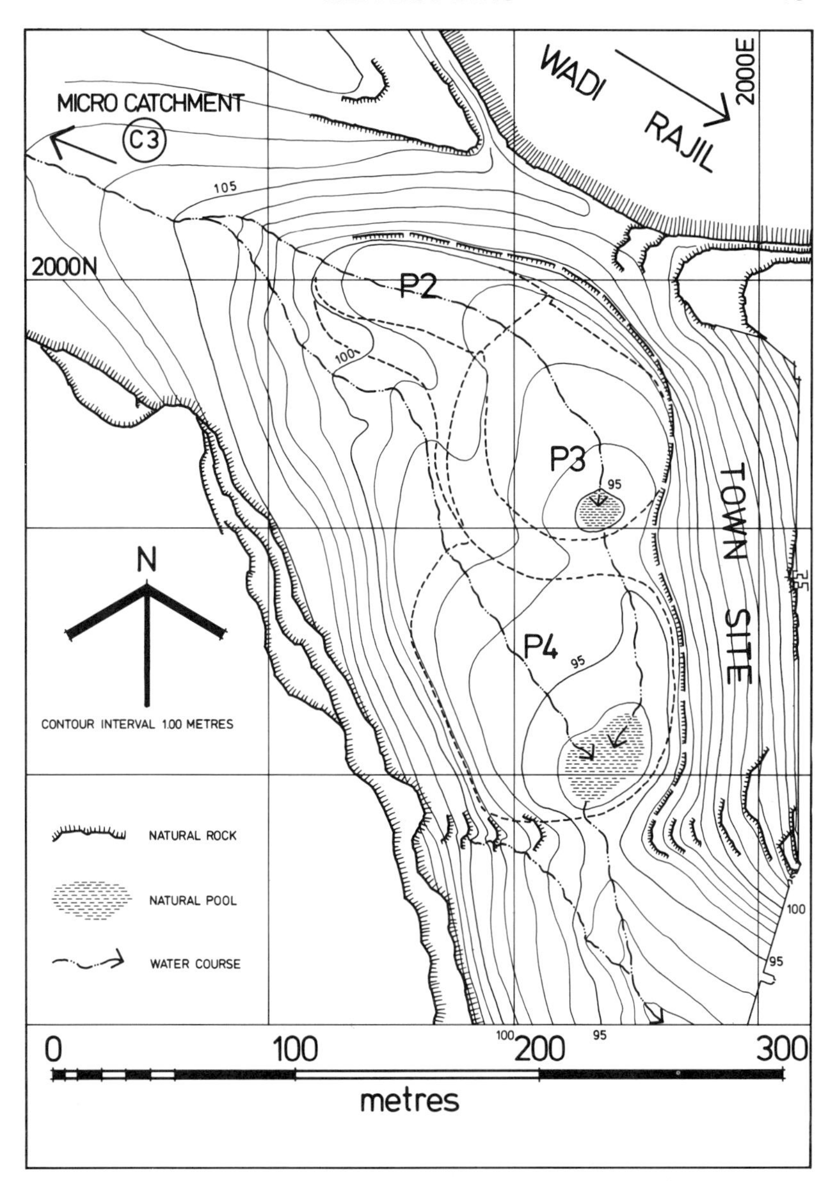

Fig. 28 Plan: reconstruction of topography at Jawa (P2, 3, 4)

people: a figure that corresponds with the population estimate of the Jawaite 'congregation' (see Chapter 15).

The annual water yield (run-off) from the adjacent land (micro-catchment) during the urban stages at Jawa derives from three linked areas (figure 71 : C1–C3). At this early stage, the bedouin one, we assume that no artificial structures were added and that only the micro-catchment nearest Jawa drained naturally into the two pools. The average water yield has been calculated as about 17,000 m^3 per year. If we weigh this amount according to probable percentage of the total rainfall per month from October to May, deduct similarly weighted monthly evaporation and seepage losses as well as the 500 m^3 that we have allotted to human and animal consumption, we may plot the end-of-month storage figures to create a very poignant curve which represents one very critical aspect of Jawa's hydrological reality (figure 29 : i).

By March the maximum storage could be about 9000 m^3, but by the end of August only 600 m^3 are left. By mid-September – before a chance of more rain – the water level is zero.

We may conclude that micro-catchment C3, the one perhaps used by the 'Old Men' of Arabia, is insufficient for a population of *c.* 2000 people and their animals, even if everyone practised extraordinary self-control and lived by imposed choice at the most basic of subsistence levels, which is often taken as 18 m^3 of water per family of six and animals for one year. And as if this were not bad enough, one can clearly see that the two natural pools at Jawa could never have held even the insufficient March maximum of 9000 m^3. At the most the pools could hold about 1800 $m^3 \simeq$ the one at P3 700, that at P4 1100 m^3. A second curve illustrating this maximum storage potential (figure 29 : ii) shows that the system becomes useless for our Jawaite population towards the end of June.

For the 'Old Men' of Arabia this meant seasonal settlement at Jawa and this is not incongruous with their bedouin status. For the Jawaites it meant failure and a continuation of their exodus. In terms of our estimated time of arrival (ETA) this means that without building anything, even if they came in October, they could not have survived as a resident, cohesive society. But they did; and therefore they must have started building their water systems as soon as they arrived.

How much they were able to build in order to use the seasonal rains depends on when they came to Jawa. The building times required for the components of the water systems must therefore be taken into account and to these must be added the time needed to build shelters and at least some of the fortifications. We must also consider the easily missed point that only a part of the 'congregation' would work on these projects. Essential to life as they are, other subsistence pursuits could not be ignored in favour of water alone. Food production and animal husbandry were as vital, especially during these fragile first days. Furthermore, some of the women and all of the small children and the elderly (if any had survived) have to be excluded from the labour force. Of the 2000 or so perhaps only about 700 could be counted on and thus the basic water scheme at Jawa, as it will be described in detail later on, would require over sixteen months for completion (see Appendix F: System I). It follows, as was said, that only an abridged version of this scheme is possible, even if the Jawaites came in October. We are faced, in our interpretation, with having to solve an equation containing several related variables: construction time, time available, amount of run-off over time and the scale of required structures, to name just a few. This is a calculation of a very hypothetical kind, but precisely the same as the one facing the Jawaites over 5000 years ago.

To solve this equation, the other side of which is success at Jawa during the fourth millennium, let it be assumed first of all that the Jawaites had a plan; that, as proposed, they had the technology. It follows that they understood the concept of planned compromise – a theme that will reappear later on in the story – and likewise that they could estimate work-time-achievement much in the manner of modern construction procedure and quantity surveying. They intended to link all three of the micro-catchments by building a canal between C2 and C1. They also planned – if time allowed and if they arrived in time – to link wadi Rajil, the macro-catchment, to their storage area. The order in which these elements were built again depended on time. Taking into account a rough work-time estimate, the canal in question represents 0·3 months, the largest pool (P4) 2·7 months and the balance, if any, could then be devoted to a diversion dam in wadi Rajil which would take advantage of one or more of the seasonal floods. The stepped line in figure 29 illustrates this graphically, the slope representing storage capacity of pool P4 against building time.

Applying this to various estimated times of arrival one can derive five hypothetical situations which may demonstrate events during the first settlement of Jawa (figure 29):

1 *ETA end-January*. The natural pools are full (1800 m^3) and micro-catchment C3 serves to recharge the pools while the canal linking it to C2 is built: by the end of February the work is finished and thence the curve reaches a high storage in April of *c*. 7000 m^3, but 'crashes' below zero by mid-September.

2 *ETA end-December*. The natural pools are full and work linking micro-catchments C3 and C2 commences but storage potential in pool P4 is not raised enough to use all of the yield from C3 over the next month; thus the storage at the end of January is still only 1800 m^3; throughout February the dam of pool P4 is high enough to take the new yield (C3+C2) and the curve reaches a high of *c*. 10,000 m^3 in April and a low at the end of the dry season of just below 300 m^3; the margin is very small indeed and if conditions in those days were even vaguely as they are today the high chance of rainfall fluctuation – especially towards the end of the wet season – could easily have resulted in a disaster and abandonment.

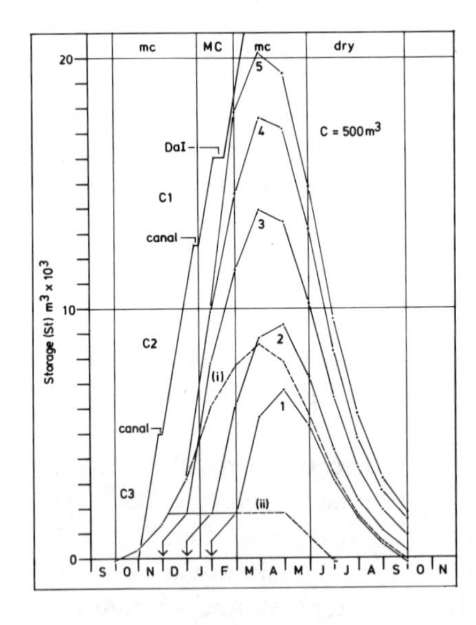

Fig. 29 Graph: situations (1)–(5) water balance

3 *ETA end-November*. Micro-catchments C3 and C2 are linked and pool P4 is sufficiently advanced in construction so that by January all of the run-off yield can be used; work would continue apace with increasing rainfall until a high storage of *c*. 14,000 m^3 is reached in

April; the low in this case would be *c.* 900 m^3 by mid-September: still a very small increase in the face of probable catastrophe; no use could be made of wadi Rajil since the time needed to build even a very small diversion dam would be too long to be finished by the probable flood period. We must therefore advance the ETA by a further month.

4 *ETA end-October.* All micro-catchments are linked, pool P4 is built to one-half capacity by January allowing all of the combined catchment yield to be used; by the end of the month P4 capacity would reach the three-quarters mark and by April the high would be about 18,000 m^3, the low in September *c.* 1500 m^3: a much more reasonable margin.

5 *ETA end-October.* This is the same as (4) above but a small diversion dam is built in wadi Rajil by the end of January diverting a modest amount of water just under 4000 m^3; the high of 22,000 m^3 comes in March – by which time pool P4 is finished – and the low at the end of September would be *c.* 2000 m^3.

From these hypothetical calculations we may conclude that the probable ETA of the Jawaites at Jawa was close to the end of October and that they might even have built a small diversion dam in wadi Rajil during their first season at the site. This means that all of the major functioning parts of the water system had been explored within a mere three months or so: that Jawa was a feasible life-support area and that after January and certainly March work could be devoted to other projects.

Now enters a broader, human factor. It has been said that the 'Old Men' of Arabia probably knew and used Jawa as a seasonal watering station. This, so far, has been defined as the logical extended period of first contact between the urbanite Jawaites and the indigenous bedouin. That there would have been some conflict is obvious. That this conflict would become serious is as inevitable as the approaching dry season.

By June the bedouins would witness a miracle. Water, up to 10,000 m^3 of it, stood for the first time in their conscious memory in the pools at Jawa. The incomprehensible toil of these new people at Jawa, these strangers from the north and another world, had achieved the impossible and bedouin would come from all parts of the Black Desert to see this phenomenon – and to benefit from it if they could. Paradoxically, the roles were reversed. The bedraggled masses that they had watched entering into their land a mere seven months before, the hungry, thirsty aliens who had begged and fought for water, now held the greatest wealth of any desert. The Jawaites owned water in the dry season.

It is pretty obvious what the next construction project would have to be at Jawa once the first and crucial life-support systems had been installed successfully. Wealth is coveted and must be protected. The high black walls, the fortifications of Jawa, bear witness to this most depressing attribute of urbanism.

IV

PLANNING, ARCHITECTURE AND COMPROMISE

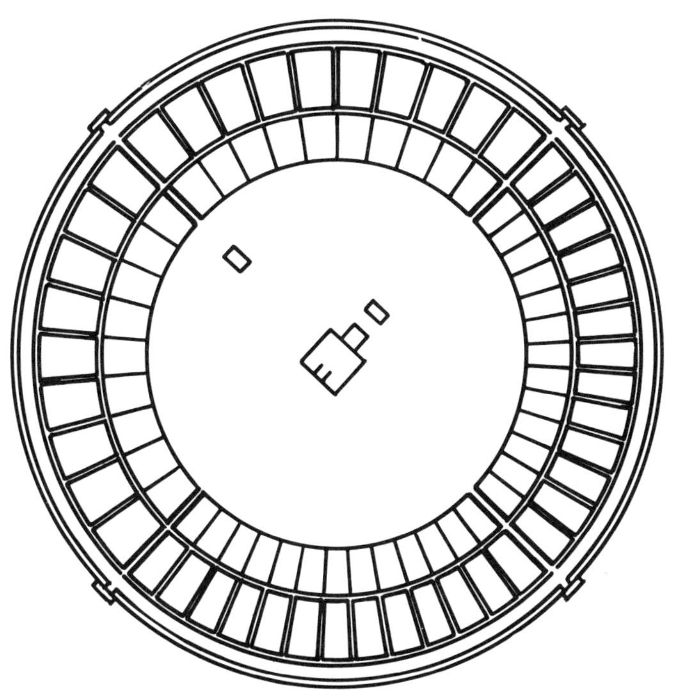

The ideal round city of Baghdad (AD 762–6): an ideal that failed after several compromises

10

THE CONTRACT

The Jawaites were in contact with the indigenous population of the Black Desert from the time they entered the region and while the first shelters and life support systems were being erected at the site. The very nature of nomadic life, dispersed and decentralized, would preclude major skirmishes between the two quite different groups. Thus, whatever the aboriginal intentions might have been – and one should not think in terms of one unified tribal 'state' – the so-called tactical moment had passed. Once Jawa was founded and settlement had begun a period of diplomatic relations ensued as within a few weeks the news of this 'invasion' travelled outwards from Jawa to the edges of the basalt in all directions and desert folk came to see these strange newcomers. The bedouin would arrive with mixed intentions, perhaps none more sinister than curiosity. Whether there was a battle or merely the threat of one, as imagined before Damascus, the heavy fortifications soon to be built at the site are more than a conditioned urban architectural reflex. Perhaps after a few indecisive clashes the situation settled down to intermittent raids, occasional negotiations and – not entirely without modern parallels – a kind of war of attrition partly relieved by peace overtures and, in general, unilateral agreements with some sector of the bedouin population.

We now come to the next stage in the development of Jawa, when the form of the permanent settlement was created and, more significantly in terms of prehistory, its form was planned and decided on. The idea of a town that the Jawaites had carried with them as their urban legacy was now to be realized; but this was to be on Jawa's terms, that included coexistence with the local population. The form of the new town was therefore influenced by the total environment: the physical as well as the psychological and

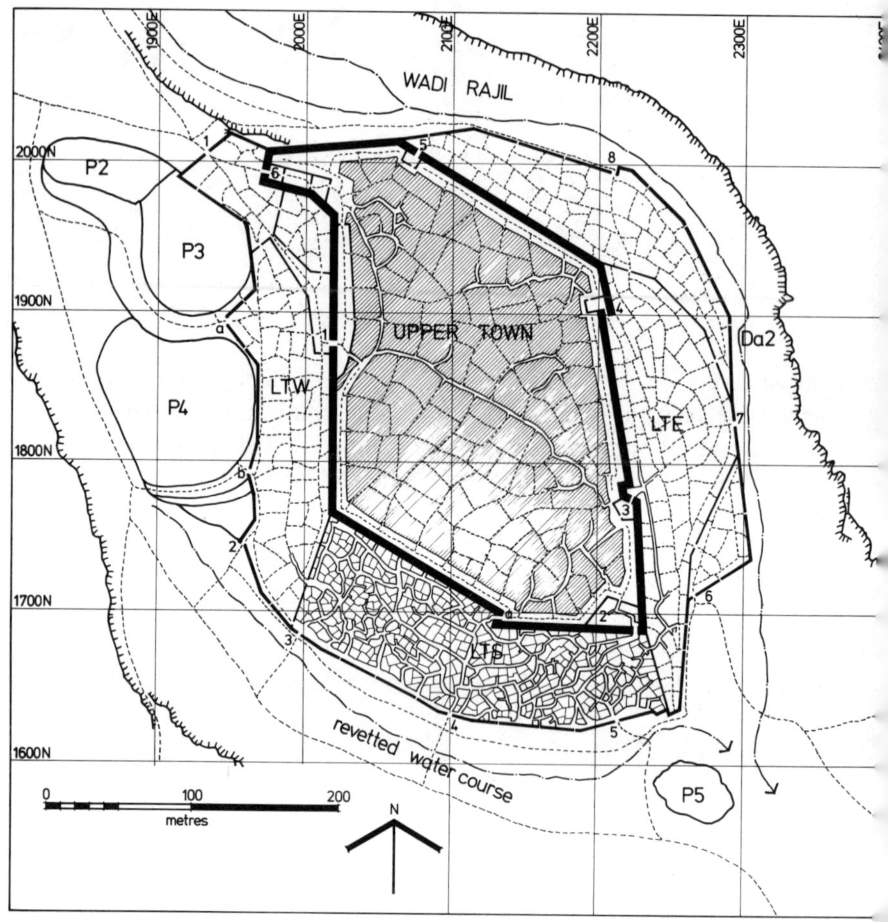

Fig. 30 Plan: reconstruction of Jawa in phase 2

the presence of a native population may have been a major factor in the new design from the very beginning.

I stress planning because it is one of several criteria of urbanism and an ambiguous term; as are city, town and village when applied to the ancient world, particularly the prehistorical and pre-literate. It has been argued by the proponents of urban revolution that civilization is the product of the city, that both are synonymous with literacy and that planning affected some of the early city's formal aspects. The major parts, however, grew like the village: organically and haphazardly in the pre-Classical world despite planning ideas (that existed) and intentions (that failed). The source of ambiguity lies partly in comparing a modern ideal of the planned city – humanized, beautiful and successful – induced by Classical models of elitist democracy and literature, with an ancient archaeological reality. The fact is often obscured that this ideal, this utopia, has never really worked.

Planning and the greater concept, the existence of the city, are therefore much older than the written word; Jawa's transmigrant urbanism probably stems from a developed civilization beyond the Black Desert. But whether we may call Jawa a city is a controvertible question; although perhaps less so than a recent reaction against the literate bias in which the status of City is conferred upon Neolithic villages like Beidha in southern Jordan (Mellaart 1975). So far as Jawa is concerned, titular protocol is, perhaps, unimportant; the idea of planning, however, is not and I take that concept to encompass what it ought to imply: intelligent and humane compromise of ideas and ideals in the real non-utopian world that includes the mundane technicalities of fortification, water supply, sewage disposal, communication, shelter – all the fundamental necessary physical aspects of human settlement – with the further reality of potential failure, under-achievement and disaster. In the case of Jawa the settlement did not grow organically nor evolve slowly and gradually without direction.

The components of the plan that the Jawaites were now to realize concerned first of all permanent domestic architecture for the 'congregation'. With this came the siting of any specialized areas such as workshops, production sites such as kilns and facilities for food storage. Next, control would be exerted over access routes into and within the settlement, which was related intimately to the general concept of shelter: protection not from nature so much as from fellow man. The Jawaite domestic sector was protected by massive walls enclosing the highest ground available at the site. This is a clear indication of a need that can only refer to the indigenous population and not, as is the case in other urbanized lands, other fortified towns, established trade networks, mutual fear, jealousy and hence potential inter-urban warfare. So far as is known other towns like Jawa did not exist in the Black Desert. Construction of the fortifications began not long after the first campfires had cooled, just over three months after the Jawaites' arrival. These almost monolithic works were therefore a reaction to an immediate local threat and with their construction yet another paradoxical situation was created whereby the new town, carefully planned and built, had within its design from the very beginning seeds of isolationism and obsolescence. It was quarantined on its hilltop and enveloped by high black walls like a prison. From the start it was cut off from its own water supply and stranded alone in the middle of an alien nomadic culture in a hostile landscape. In short, the Jawaites were probably beginning to develop the typical encapsulated psychology of a foreign intrusion concentrating itself in a large, intellectually as well as technically brilliant ghetto.

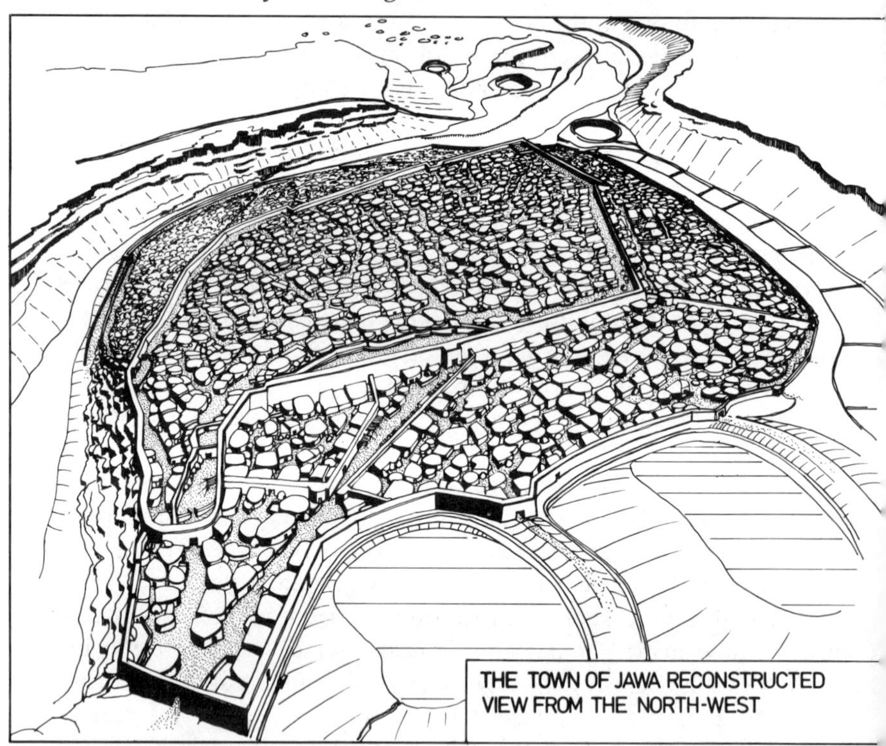

Fig. 31 Perspective reconstruction of Jawa

To alleviate this situation certain steps were taken which can be measured and attributed to further thought and planning from within Jawa: notably access to its domestic citadel. Reference to the illustrations will show that points of entry were strictly limited and it seems that only four – possibly only three – gates existed at this stage. Gate UT6 was rebuilt later and may not have been in use now. The major gate, UT1, centrally located in the long straight west wall, was approached along the main access route from the flatter land to the west. The southern and eastern sectors of the upper town each had a gate (UT2 and UT3) giving onto the lower slopes of the site towards wadi Rajil.

The development of the three external slopes below the upper town walls further underlines the almost necessary topographical isolation of the Jawaite core while at the same time demonstrating certain measures taken to relieve resultant problems. Very soon after the construction of the upper fortifications three fortified sectors were laid out to the west, south and east. These suburbs, as insinuated, may well have been meant for the non-Jawaite settlement of the site. The only alternative interpretation of these lower quarters concerns either the idea of population growth within the upper Jawaite town or social stratification within the immigrant urbanite community. A Jawaite population explosion may be ruled out for two reasons: the time span of occupation is far too short – perhaps merely a part of one short generation – and spatially the lower quarters represent a population rise of well over 100 per cent. Total absence of architectural and zonal specialization in the upper enclave tends to preclude formal social stratification among the Jawaites. But the overall shape and division, as well as the location of Jawa's fortified domestic sectors, does point to stratification based on ethnic, cultural and sub-racial grounds.

There is some proof that access to the upper town was given special attention, almost as if to safeguard ethnic purity within it. This is visible in the aerial photographs that show a walled ramp along the western access track to gate UT1, not only indicating controlled access to the water storage areas and the fields beyond, but also segregation of the lower town in the west quarter. The concept of controlled access, with all its neurotic implications and the clear reference to separation of domestic sectors, is normal to the architecture of Jawa and a part of the general plan which developed in a series of compromises. Of the lower-town quarters the western one is the most causative since it lies between the upper town and the water supply.

Access to the lower-town quarters, as these developed, is much more typical of what was to become the normal urban design in Palestine during the next millennium, after the fall of Jawa. The apparent order at Jawa is another indication of conscious planning that was directed from the upper town. Each of the lower quarters was given at least three gates, roughly equidistant from each other, very similar in design and supplemented by many simple posterns. For most of the time such a profusion of gates would serve simply in a domestic capacity, allowing relatively free movement through the fortified perimeter. In times of siege they could all be closed very quickly. Because of the remarkable state of preservation at Jawa we are able to understand the entire scheme as a planned one; and as one of the earliest and almost complete examples of the nucleated town found to date.

All of this, form and construction, came into being suddenly, yet it was less of a miracle than plain hard work that was deliberately controlled and guided according – as it had to be – to a flexible plan. Let us therefore look at the sheer scale of the undertaking in purely physical terms.

The total built-up area of Jawa at this stage (phase 2) was in excess of 100,000 m^2: roughly 10 hectares or just under 30 acres. Jawa was larger than most of the later fortified towns of Early Bronze Age Palestine and yet no more than the equivalent area of twelve average modern city blocks.

Jawa's building material requirements were staggering (see Appendix F). The total length of the fortifications is about 2700 m of which just over 1000 m belong to the upper town, the balance to the external and internal walls of the three lower quarters. For the upper town this represents roughly 12,000 tons of basalt; for the lower town about 36,000 tons: a total of 48,000 tons of rock that included boulders measuring nearly 2 m in length. If we now add the domestic structures within these walls the final estimate is about 98,000 tons of basalt, 37,000 tons of mud for bricks, 22,000 tons of mud plaster and about 2000 m of heavy timbers for roof supports and joists and 42,000 m of smaller scantlings for other structural woodwork. This excludes brick superstructure on the fortifications, rendering these in mudplaster, industrial installations, domestic structures apart from storage bins, incidental structures in lanes or streets, paving in bins, straw binder for the bricks, reeds, lintels, door jambs and so on. Apart from these materials there would be those required for the water systems. The total then is just under 300,000 tons of raw materials that had to be obtained, processed, transported, worked, bound and hammered into place by hand.

But how long a time and how many workers would be needed? How long a time could be afforded? And, if a strict timetable was imposed by the environment as proposed, were there enough Jawaites to cope with this ambitious plan?

In order to quantify the task of building Jawa we must first of all realize that construction, especially on a monumental scale, is a sequential matter. The finished product does not spring into being, as is often the impression from archaeological accounts. The various elements that make up the physical structure of Jawa have to be valued in terms of materials (weight), relative importance, time available and energy requirements, among others. Calculations are summarized in Appendix F.

Unassisted, the Jawaites could build everything that we see as a ruin today in just over three years, slowly improving the catastrophe margin of the water systems throughout the period. However, this does not explain the three lower quarters; nor does it reflect what we choose to call Jawa's demographic reality: that the Jawaites were not alone.

Let us therefore return to the hypothesis set out in the previous chapter and explore the continuation of the story.

Bedouin pressure on Jawa was an obvious threat to security and once the basic water systems were finished (about March) work began in earnest on the upper fortifications. We may suppose that about one-half of this task was completed in three and a half months and that the most vulnerable flanks, the west and south, were now protected by the high black walls visible today. During this time work would also have continued on domestic architecture. It would be mid-June, precisely the moment at which we left the story earlier, and pool P4 would gleam in the sun as a magnet to the bedouin. Pressure on the new urban system would increase as the dry season progressed and although the fortifications and presumably the fighting skill of the Jawaites might have countered any overt action, in view of the structural evidence a compromise seems more likely. It was argued earlier that some form of agreement might have been made with the bedouin at the very beginning. Its precise form and content is not clear (naturally) but we might now suggest that under the changed and stressed conditions the Jawaites entered into a contract with the 'Old Men' of Arabia.

Bedouin wanted watering rights and Jawaites security. The latter might trade water against labour, realizing that their fortifications were still incomplete and also that the new reality demanded an upward adjustment of the water consumption rate and hence further construction in that sector of Jawa as well. The bedouin, in addition to water, might have been interested in manufactured goods that the Jawaites had brought with them from beyond the Black Desert. In general these bedouin would come for a variety of reasons – their curiosity has been noted – and some of their motives were clear and immediately formed at the sight of Jawa's water miracle. Other motives would develop through contact with the new technology and culture of the urban folk. We have here the equivalent of a very modern situation with similar potential problems: the *Gastarbeiter*, the paid and hopefully transient, virtually expendable slave labourer.

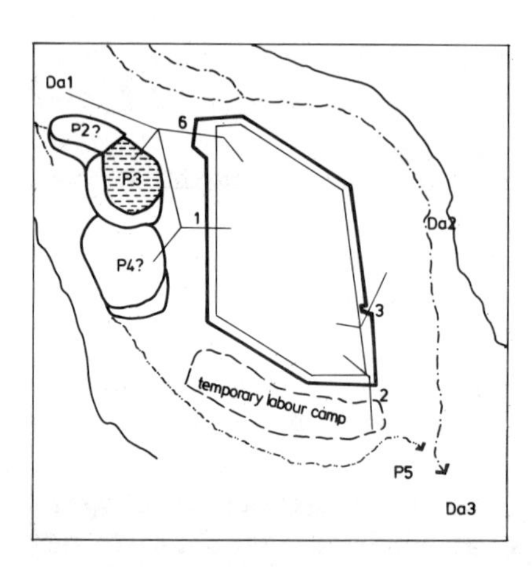

Fig. 32 Plan: Ideal Jawa

The bedouin workers would be allowed to settle in temporary shelters near the work-site, probably accompanied by their families. As work proceeded the question would arise among the Jawaites of what to do with these people when the job was done. A prehistorical Machiavelli would have stressed the fortifications now being built with the help of the bedouin whose camp lay unprotected below the south trace and the need to 'deal wisely' once the contract ended. In the lower social strata – if modern parallels are any guide – another, quite different possibility would have formed in the minds of the workers long before the end of their agreement: the idea of permanent settlement at Jawa, in a technically superior culture, rather than a return to the harder life of the Black Desert.

If this is accepted as a reasonable model we know whose ideas prevailed. The suburbs described above were built on the pattern of the upper town, on a lesser scale and without so much care, but nevertheless surrounding the nucleus of Jawa. I propose therefore that about 500 or so bedouin workers and their families (a total of *c*. 1500) began to settle at Jawa in the early summer. They would help build the additional elements of the water systems and upper fortifications and by January start on 'formalizing' their own domestic sector. Water consumption at this stage would be adjusted to a new level (see figure 94). This point of Jawa's urban history might be regarded as the first great compromise of an ideal and as is the nature of compromise would lead to the next. Modern parallels have taught that without 'draconian' measures immigration into a more comfortable environment is a one-way reaction to population pressure. It is not so very far-fetched to suggest that more bedouin came to settle and the Jawaites simply could not afford any counter measures other than further compromise. Accordingly, the water consumption rate was increased yet again around April or May and probably again after some crisis which, at this stage of the story, is still in the future. One can imagine the growth of the lower quarters as a continuous process over which some measure of planning control was exerted. Direction still came from within the black-walled brain of upper Jawa.

The urban organism of Jawa is almost completely developed but, like any living system, mortal. How long it survives depends upon many things; but if, as we are slowly demonstrating, it develops serious flaws virtually at birth, any extraordinary stress can kill it. Very negatively, one could even regard the organism as being infected by an incurable disease and the defects, in urban technical terms, as follows: access to the water sources blocked by the growing western suburb, access to other life-support systems blocked in the east and south, upper town surrounded on three sides by suburbs of dubious loyalty, the question of control over growth with regard to space in the town, the question of growth with regard to the water supply, maintenance of the water supply, flood control, control of the labour force, control of further immigration, control of the animal population and so on. One could go on and eventually list all the technical and psychological problems of urban life and class/race discrimination, of any time or place, and contemplate prophetically the plan and the reality: the idea and the growing thing which at Jawa, if one is very pessimistic, is as good as dead before it has properly been born.

II

COMPROMISE

Let us however postpone the inevitable and seek cover for a time from reality, as did the Jawaites in the fourth millennium, and consider the traditional expression of civilized neurosis.

At Jawa we find a text-book illustration of military architecture that for the first time in the study of prehistory might be completely evaluated since it is all still above the ground. In this analysis it will be seen once again how the ideal of Jawa changed and how the form of the town was altered through a series of compromises.

The choice of site is in one sense tactically correct by building the fortified town on high ground surrounded by steep cliffs on most sides. On the other hand such siting raises the question of access to the water supply that has, ideally, to be intramural if sieges are to be withstood. This was resolved because the town bordered on the reservoirs and the population could draw water under cover of whatever defensive artillery there might have been. The water source could also be defended should the need arise. Bow and arrow as well as sling and stone are effective over ranges above 100 m and all storage areas vital to urban survival are within this limit. However, it may be significant that the fire perimeter is the line of the lower town fortifications. We have already insinuated how this situation could become strategically equivocal for sociological reasons. This is further aggravated, though nothing could be done about it – even less than the social question – by the very nature of the water systems whose source of recharge had to be extramural. From a military point of view the parallel here is a matter of being not only flanked but surrounded by and also dependent on troops of mercurial fidelity who, moreover, are sub-racially, ethnically and culturally much closer to the potential enemy than to their leaders. And what is worse, all of this takes place in a static defensive position with tenuous and over-extended supply lines.

The trace of fortifications consists of long straight walls without towers and this means that flanking projectile fire could not easily be directed at attackers once they had taken the lower fortifications. The lower line was held by a potential fifth column. Only in the immediate vicinity of the gates are there any signs of projections beyond the walls. Only in a few instances does the upper fortification wall jut in or out to allow anyone to shoot along the line. The lower trace of fortifications may once have had two small rounded towers, but on the whole the trace at Jawa was no more than a high fence of urban proportions and therefore tactically weak. This much is obvious when one notes the heavily repaired breaches

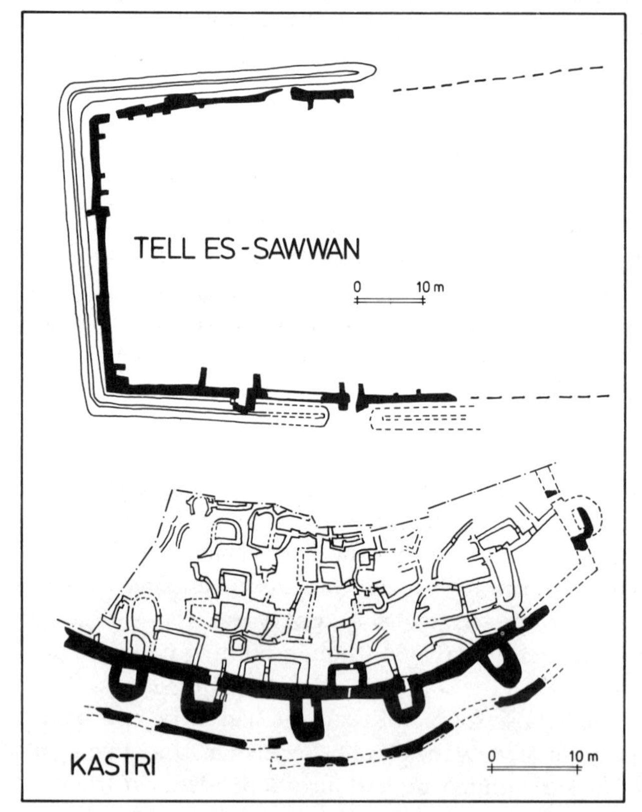

Fig. 33 Plan: (a) Tell es-Sawwan (sixth millennium) and (b) Kastri on Syros

and the dark import of this in the west line of the upper town and also in the south at the end of phase 2. Finally, although some planning control was still evident in keeping even the lower fortifications free of domestic structures, troop movement from the upper town to the lower would have been slow and further impaired by the presence there of a potentially hostile and unco-operative population. The upper-town trace then is not merely the inner line of defence, it is the last and for many reasons the only one. Its character, as it emerges, typifies the strategy inherent in Jawa's military architecture, which is completely static and not very advanced.

More parallels will be given in discussing the gates of Jawa in the following chapter. With regard to the trace of fortifications we may refer to much earlier military architecture in the Near East which shows us the very ancient roots of this technology. At Tell es-

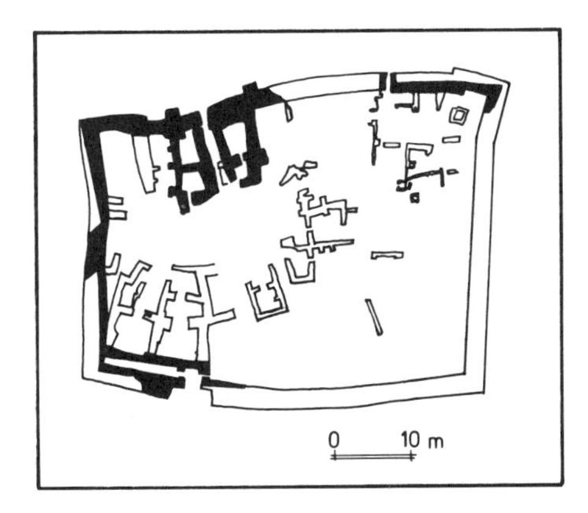

Fig. 34 Plan: Hacilar (fifth millennium)

Sawwan on the Tigris river of Mesopotamia we have a slightly more advanced trace consisting of long straight walls preceded by a ditch, where only one tower can be clearly and definitely identified – and it is actually a kind of gate. Similarly at the site of Hacilar in Anatolia, closer to Jawa in date but still substantially earlier, no towers at all exist, only a complicated gate in a 'trace' that is hardly more than a fortified farmstead. In Transjordan and Palestine regularly-spaced towers do not occur until the Early Bronze Age. Little, if anything, is known in the technical sense about fourth-millennium fortifications in Egypt and Mesopotamia, but we do know that large fortified

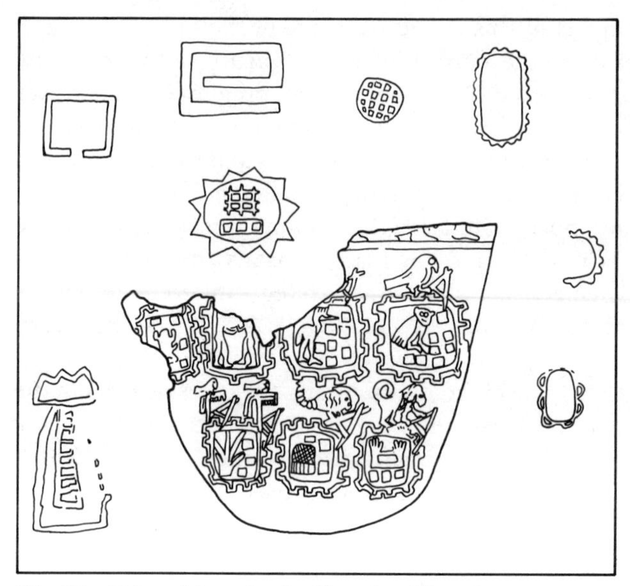

Fig. 35 Urban hieroglyphs (fourth millennium onwards)

towns must have flourished there. Hieroglyphs representing such places are known, but the projections from walls depicted in them have also been called buttresses, rather than towers. Broadly speaking Jawa occupies a median position between designs like that of Tell es-Sawwan and the third-millennium fortifications of which Kastri on Syros in the Aegean Sea is a good example. At Kastri the debt – obviously an indirect one – to Tell es-Sawwan and the emerging military architecture at that site is clear when one compares the gates. Thus the rather fugitive rounded towers in the outer fortifications of Jawa, even if they stem from non-Jawaite tradition, may have a place in the general development towards traces like Kastri.

The walls of Jawa are impressive even if they are tactically deficient. There is an obvious difference between those of the upper town and the lower, although the techniques of construction are much the same. Both are made of basalt roughly coursed with a rubble core. Only the scale is different. The upper fortifications still stand up to 6 m above the bedrock and are an average 4·5 m wide at the base. The nature of the coursing, in the absence of cut and fitted stones, produced the marked batter on both sides of these walls, the interior being made of smaller stones. The large boulders in the exterior faces would make scaling the walls very easy, unless they were plastered over with mud and maintained that way after the wet season. All along the upper trace a natural rock scarp was followed so that the inner face often required only a few courses. In this way the upper walls resemble the lower ones which are coarser and smaller but also set against a scarp or the sloping occupational debris of the previous phase so that the interior faces were only two or three courses high at the most. Their appearance was more like a long line of retaining walls which nevertheless could be defended and must have been designed with that purpose in mind. The lower walls of Jawa were also provided with relatively sophisticated gates. (For sections through the upper and lower fortifications see figures A2 and A3 in Appendix A.)

This brings us to the question of original height and other tactical considerations. The former is a simple matter since the extant remains at Jawa give us sufficient height that would be tactically sound: that is, walls in the upper town in excess of 6 m above the external ground and about half that for the lower trace. Superstructure is another general problem when dealing with early fortifications anywhere. It can be assumed that some sort of parapet was constructed, even if only a thin wall one brick wide to keep people from falling off at night. Battlements, merlons, crenels and other embellishments may have existed at one time but are no longer indicated by any architectural evidence at the site. Let it suffice that such features are well known before Jawa and have been illustrated in early hieroglyphs and even on predynastic Egyptian drawings of ships which show battlemented fighting platforms, the forerunners of ships' castles.

Jawa has more gates than have been found among any of the fortified towns of Early Bronze Age Palestine, towns which for various reasons form the most comprehensive corpus of early urban fortifications known anywhere in the world. Jawa indeed has more gates than any prehistoric settlement yet discovered. The recording and analysis of these gates has accordingly produced a manual of urban military architecture for the later fourth millennium which has some relevance in terms of precedent to later development in the Near East. Because of their state of preservation (ironically, owing to

Jawa's failure) we can study more than structure and style, but also concept, planning and the overall systematic design within the greater architectural scheme. Seldom can so complete a picture be presented in a town site so early in man's history.

But let us first turn to the plan or the form of ideal Jawa. We must realize of course that it may be difficult to recognize this in its pure form because that never existed. The ideal suffered one compromise after the next almost as soon as it was conceived. Such changes must therefore be stripped from what can be seen at the site now.

The ideal Jawa may well have been a purely Jawaite town without suburbs. Although undoubtedly the town was built with the help of local bedouin labour this contract was meant to be terminated and it was hoped that the locals would go away once the job was done. But since this was not to be the case (Compromise 1) the new, altered form of Jawa was still at least to be controlled from the Jawaite core and arranged in an orderly way. The design of the lower quarters repeated the 'excellence' of the upper town, perhaps at a humbler level, just like modern suburban extensions copy the spirit of a once patrician core. In this perhaps lies a further aspect of the relationship between the two groups at Jawa. Whoever lived in the suburbs was powerful enough not only to be allowed to settle in the first place but also to be given quite similar domestic architecture – not just shanty-town shacks – and to affect the core of the town, the controlling founder centre. This last point is illustrated in amendments to the gates of the upper town, departing from the ideal plan in a very telling way.

The plan was changed to include suburban quarters below the upper town and we should recall the problems cited earlier, both immediate and potential. This compromised plan is shown here and bears some resemblance to later town plans that have been found cut into mud and stone panels from Mesopotamia. In those the plan was less concerned with domestic arrangements than with the general framework of walls, gates and major access routes.

Jawa was to consist of an upper town as the directive nucleus with six to seven gates, a wide ring-road just inside the fortifications and within that the domestic quarters of the Jawaites. The gates, originally only four or three, were probably intended more for access to extra-urban space than communication with the suburbs. Thus in the east the original gate (UT3) now led to the middle of the

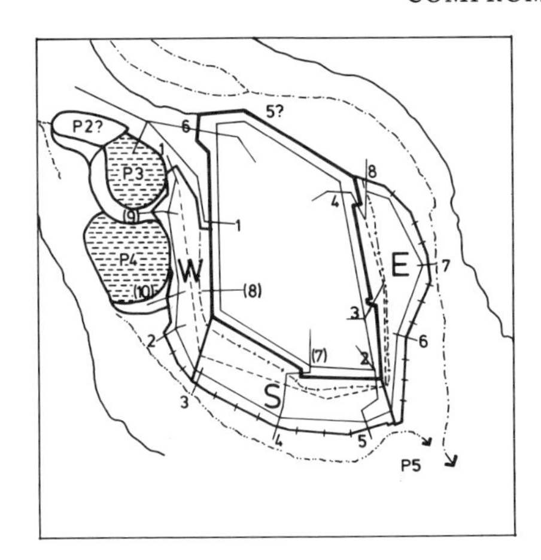

Fig. 36 Plan: Jawa Compromise 1

eastern suburb and perhaps gate UT4 was made to gain more direct access to the north and wadi Rajil. Similarly gate UT2 might now have been supplemented by gate UT(7). The western quarter was the most problematic and indicative of the compromise that had to be. That quarter lay next to the water systems' storage area.

It has been stressed that we do not have archaeological proof in the proper sense but are able to interpret much because nearly all of Jawa is visible on the ground. It can be reasoned, therefore, that this west part of the town might, in its development, betray more of the kind of struggle that took place.

During the second stage of compromise the Jawaites might first have tried to keep the area open and failing that at least controlled. The ramp up to gate UT1 is a symptom of this, as is the (later) breach farther south which may actually obliterate traces of an additional gate UT(8) similar to the arrangement in the south and east flanks of the town. More important now than the signs of an orderly geometric pattern, however, is the situation north of gate UT1, up to the north-west corner of Jawa and the site of gate UT6 whose form at this stage was probably different than at the end of Jawa. This much is clear from the aerial photographs. The area between the two gates was meant to be open so that pool P3 could serve the water needs of the upper town without obstructions. Pool P4 was already serving the lower western quarter.

The plan – in theory and for a time perhaps even in practice – was for two gates leading from the upper town into each of the three lower quarters linking the two circuits of fortifications as well as the two main lines of communication in concentric ring-roads, themselves linked by cross streets of smaller size (less planning control) and possibly supplemented by a median ring-road in the lower town areas that passed through the internal fortifications at various gates, some of which are still visible from the air. This network then branched again to three gates in each of the lower quarters and to numerous posterns. Concentricity is broken only in the north-west for reasons already noted. There are – significantly – no posterns in the upper-town fortifications.

This organism grew and continued the process of enveloping its controlling nucleus like a mechanized armoured column split in two to outflank an enemy position, eventually to surround him completely in the classic pincer movement. It was a movement that may have begun in a temporary labourers' camp to the south of the new town, a constant pressure that had to be controlled even if it could no longer be stopped. Inexorably this led to further compromise towards the latter part of this phase.

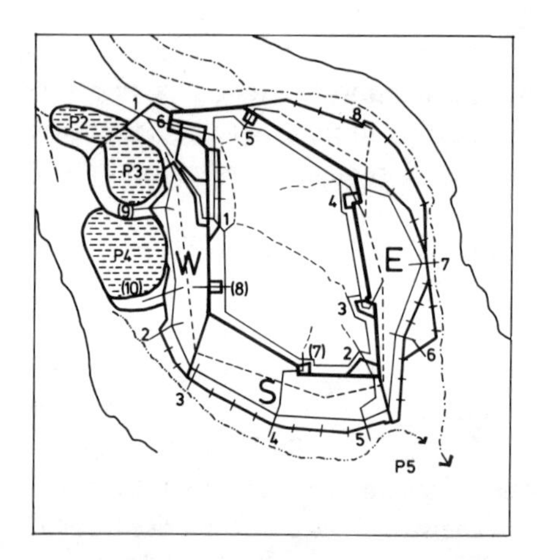

Fig. 37 Plan: Jawa Compromise 2

The second compromised ideal of Jawa is reached as a result of more suburban growth. The lower-town quarters expanded. In the east two new sectors appeared, each walled by retaining structures and provided with gates along a new line of trace that now stretched to the topographical limits of the site, up to the rim of wadi Rajil. Yet another gate had to be added in the upper town (UT5) and the same probably happened in the south where this process was to be repeated a little later on. But most critical was the west sector. There the lower town now definitely spread up to the cliffs facing wadi Rajil, outflanking both gates UT1 and UT6. The upper town, but for a narrow strip of steep cliff in the north, was now totally encompassed.

Reaction within the upper town (Compromise 2) is clear and perhaps justifiably paranoid. Gate UT6 was extended and complicated westwards. Gate UT1 was secured internally by a second line of defence and an internal gate obviously designed to control ingress from the western lower town. On the military level this is the equivalent of a second line of defence erected hurriedly, almost like a field-work, once the major trace is threatened with total collapse. Such works, barring miracles, normally preface the end of static defence systems.

Other gates in Jawa's upper town were similarly reinforced from within and the pattern of the ideal was broken, as was the line of the upper ring-road. Jawa had reached a spatial, technical and psychological watershed; probably a demographical one as well.

The military parallels cited above are quite suitable here since the story of Jawa now proceeds much like a losing battle. It is a struggle that is essentially like trench warfare in which the protagonists face each other across their static defences in an obscene intimacy that permits occasional peaceful communication as well as murder. The Jawaites in their stronghold, which is what the upper town had become, were completely enveloped and cut off should conditions ever deteriorate. One compromise after another had left them like rats in a corner: paranoid, introverted, neurotic and probably vicious; a prehistoric version of stressed ghetto mentality.

12

THE GATES OF JAWA

Despite certain failures, the architectural achievement of the Jawaites is remarkable. This is as true of the scope of their task as it is of design and we must now look at the structure of the gates at the site which protected the Jawaites from the lower quarters at the end of phase 2.

Two examples will be described as representative of both fortification lines. Gate UT1 is the major gate of the upper town and

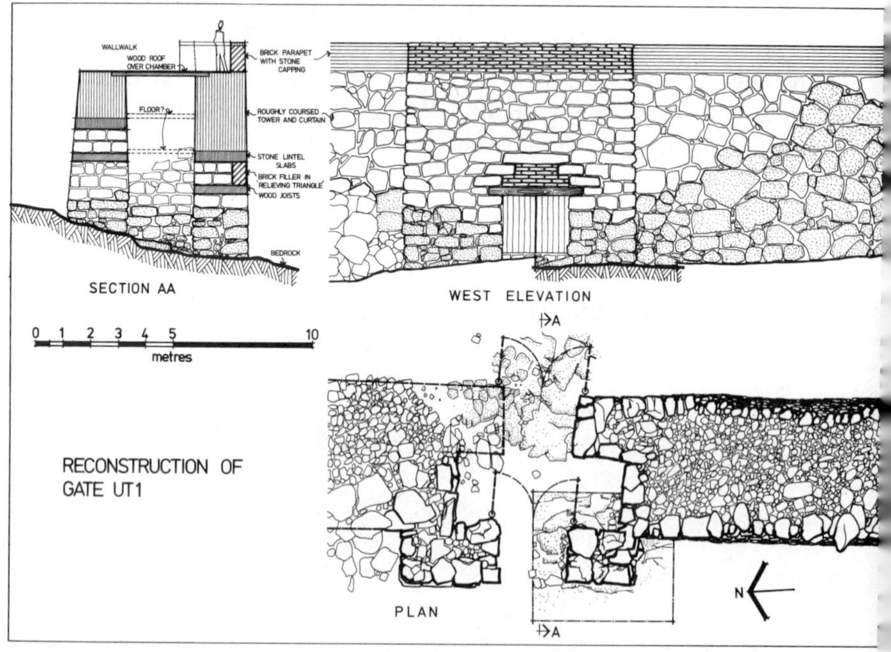

Fig. 38 Plan, section and elevation of gate UT1

gate LT4 the best preserved of the lower.

UT1 always was the main gate at Jawa, even when its approaches were threatened by suburban sprawl, and that is why it has survived as the best of them all. Its structure and design indicate a high standard, as one would expect of a main gate, with all the inherent symbolism of the major entrance to Fortress Jawa. It is therefore possible to restore most of its original appearance.

The lower parts of the gate were bonded into the curtains on either side and differentiated only in the relatively regular coursing – owing to the choice of building stone – much in the manner of quoins in otherwise roughly-built masonry. As noted above, the upper line of fortifications was built along a natural rock scarp and this entailed some complication in whatever roofing scheme there was over the gate. The gate itself consisted of a single rectangular chamber whose long axis is parallel to the line of trace. The chamber is flanked by heavy stone jambs or internal buttresses. It is very likely that two sets of double-leaf doors were used. Although we found no socket stones *in situ* here we did find them in gate LT4. So far the upper-town gate is quite a simple structure designed to control entry, should the need arise, but also to permit normal traffic. At the same time it would offer some shelter to guards and we must accept that the chamber over the entrance was roofed in some way and ask two questions. How was the requisite space between the jambs or buttresses spanned? And, how far up did the stone-work reach? To answer these questions absolutely is impossible because nothing is left above the fifth course within the gate. However, using building materials readily available at the site one can suggest three possible reconstructions.

Most basically, the stone could have risen all the way up to wall-walk level. This would impose a very heavy load on whatever spanned the doorways. Secondly, the section immediately above the doors could have been made of brick from the level of the door lintels upwards, thus reducing the load. Finally a compromise – as usual at Jawa – and stonework could have risen all the way up on either side of the doors with bricks in between.

Whatever materials one chooses for over the tower there was some considerable load above the doors. We may consider a series of possibilities for reconstruction, all within the technical capacities of the Jawaites. There could simply have been a wooden lintel, or even

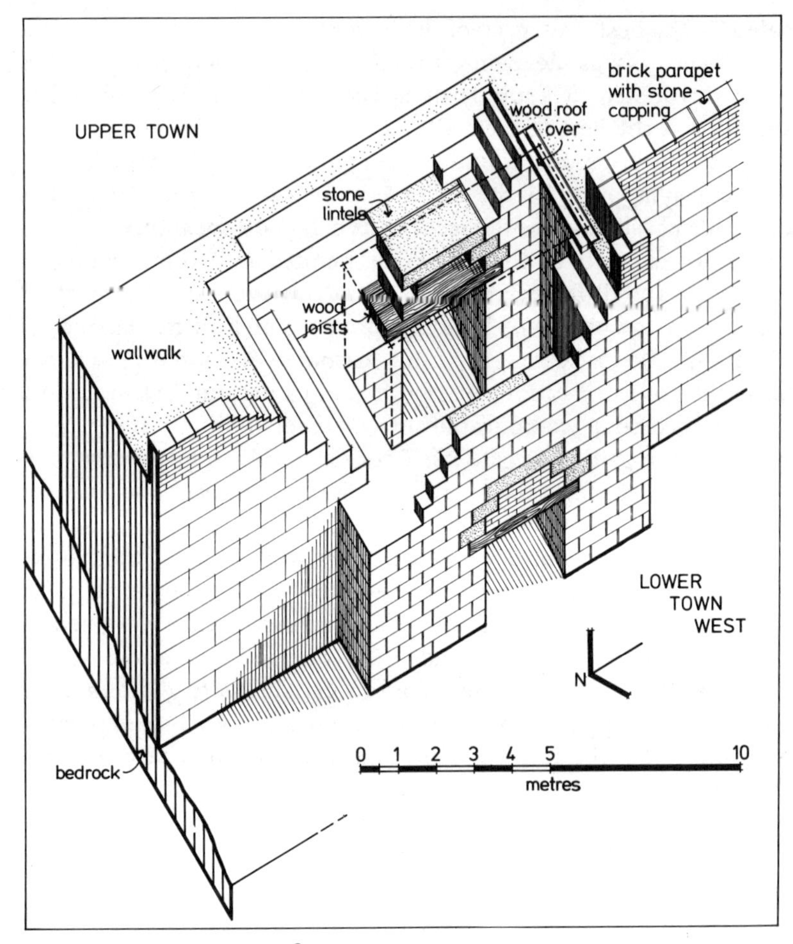

Fig. 39 Reconstruction of gate UT1

one of stone providing the proper length and size were available. A more complicated technique might have taken the form of a relieving triangle, either of wood or stone, as illustrated, which would disperse the load towards the jambs on either side of the doors. Or one could suggest a version of the corbelling technique that was used at Jawa during the Middle Bronze Age at the Citadel. This raises a rather interesting point: the builders of the later structures might have passed beneath the corbelled entrances of Jawa's ancient main gate and been influenced by what they saw – a not entirely impossible suggestion, but one that would tend to disrupt normal theories of architectural development.

Figure 39 depicts the most advanced of these possibilities as an exercise in applied theoretical reconstruction that is quite plausible given Jawa's technical brilliance elsewhere at the site.

Since doors existed (double-leaf) we may assume that they were meant to shut properly; something that is often overlooked in architectural reconstructions. A lintel of some kind would therefore have existed and its form, once one accepts the principle of corbelling, is shown here. It is a version of the relieving triangle which is the ancestor of the tympanum of Roman and Medieval architecture; it is also precisely the technique used in the tomb

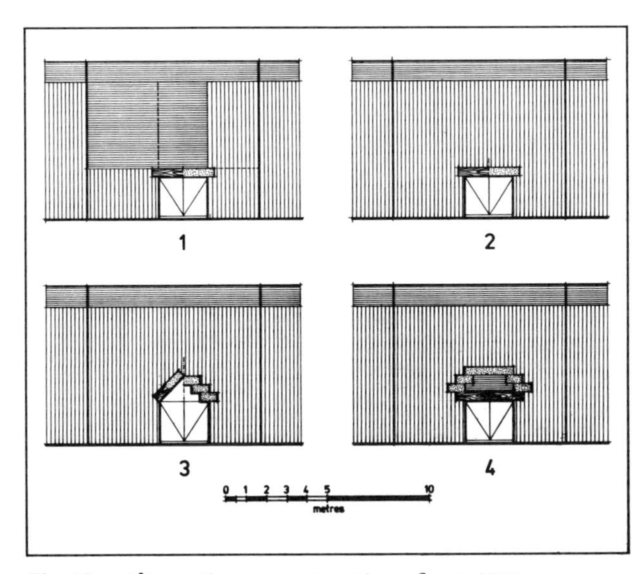

Fig. 40 Alternative reconstruction of gate UT1

chambers of the Great Pyramids of Egypt built several centuries after Jawa and reappears in pre-Classical Greek tombs such as the Treasury of Atrius at Mycene in the Peloponnesus. Above such a roof any height of superstructure can be reconstructed which may have given wall troops additional shelter. Technically there is no problem. One could add several storeys and they would stand. This was indeed often done in much later military architecture, sometimes with the embarrassing result that the enemy at ground level was almost beyond the range of available artillery.

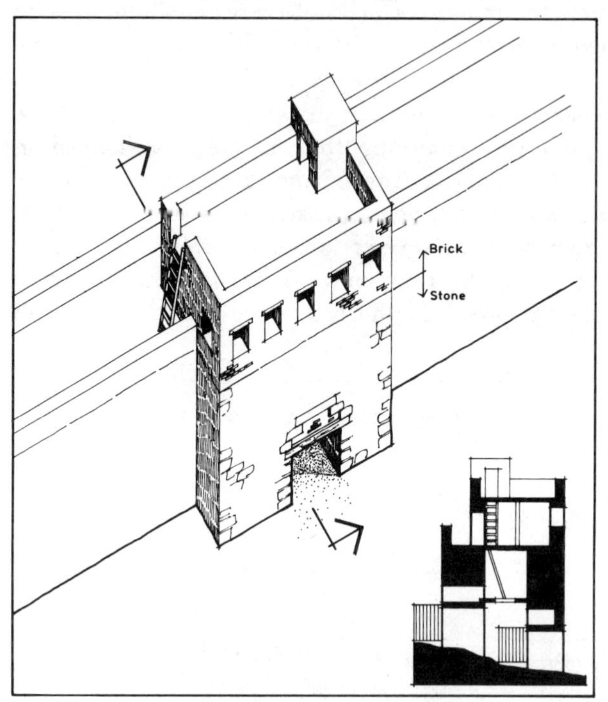

Fig. 41 Another reconstruction of gate UT1

The lower fortifications were of similar design and thus the products of the same planning intelligence. Only minor differences in finish and scale existed. Gate LT4 has two oblong chambers and thus a set of three internal buttresses. Socket stones were found in place as shown in the plan here and substantiate the erection of three separate doorways along the upward-sloping surface into the lower town. This surface was provided with a series of stone steps. The superstructure was probably similar to that of the upper-town gates, the only difference being that the original height would have been less because the lower defences were really more retaining walls than a series of proper curtains in the strictest sense. The wall on either side of the gate is interesting. It is no less than a true casement made up of an outer and inner continuous wall with cross walls at intervals. The resulting structural matrix was filled with rubble and

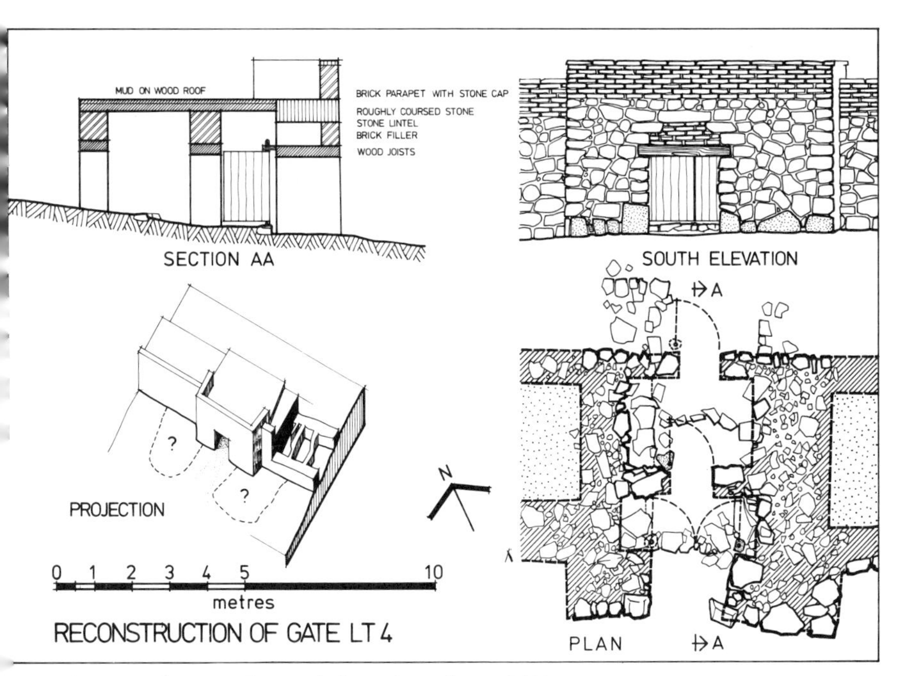

Fig. 42 Plan, section and elevation of gate LT4

earth. Such walls do not appear commonly in the fortifications of the Near East until over a millennium later, although a few rather vague examples are known in Palestine from the Early Bronze Age.

These two gates, single- and double-chambered, were repeated throughout the fortifications at Jawa with but minor variations resulting mostly from adaptation to the terrain: another quite advanced principle. Thus, further along the southern trace, gate LT5 is related to gate UT1 (single-chamber) but was later made into a more complex structure by adding fortifications on the west side. This caused a slightly oblique approach rather like the ramp to gate UT1, an indirect-access gateway. Similarly gate UT3 in the upper line has a double-chambered direct-access design but is sited in such a way that the ground in front forces an indirect lateral approach. Figure 44 includes a number of reconstructions. Note gate UT1 and the second compromised ideal with internal fortifications added. Note also that these gates are the closest one comes to projecting wall towers at Jawa; a point discussed earlier in relation to the tactical level of Jawa's military architecture.

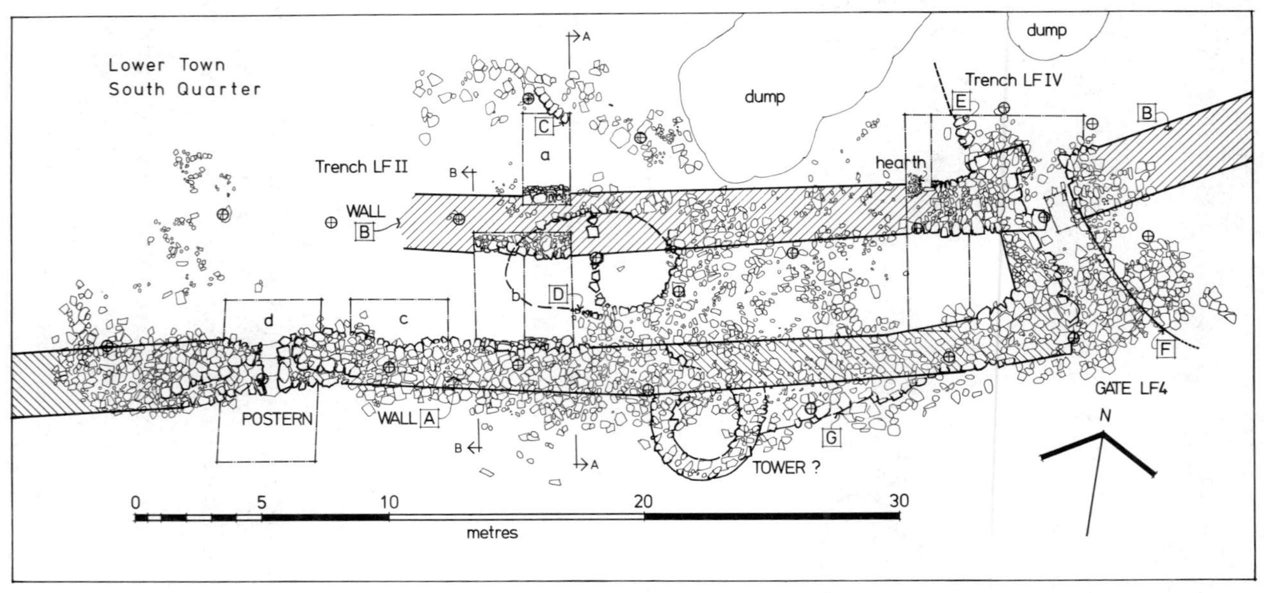

Fig. 43 Plan : area LF trenches II and IV

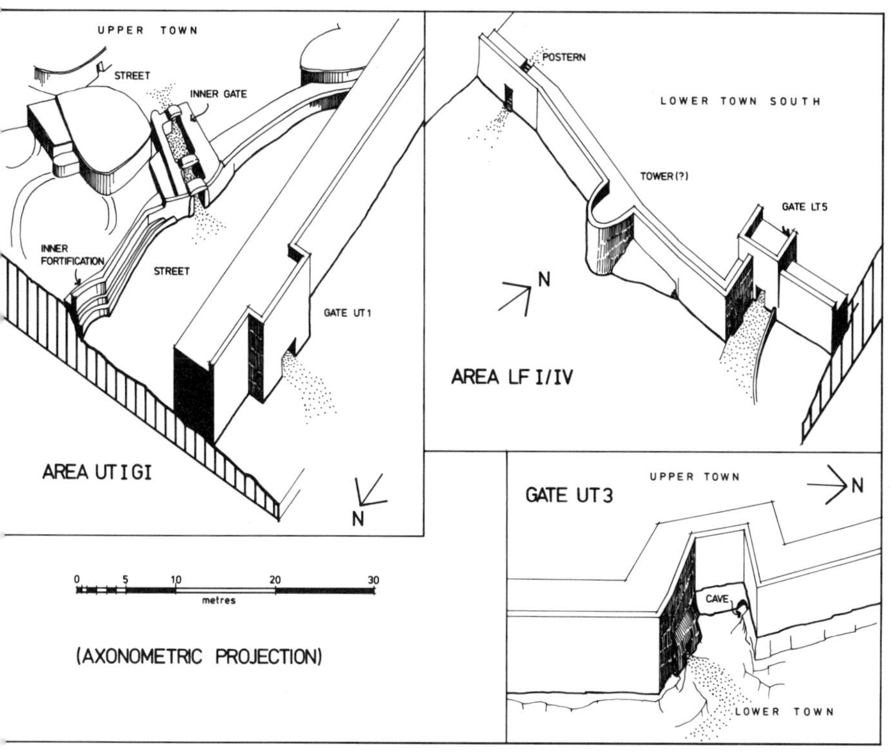

Fig. 44 Reconstruction of the gates of Jawa

This brings us to a more general appreciation of Jawa's gates, which through their abundance allow more than a casual comment regarding the development of this branch of military architecture in the Near East.

A gate is simply a hole in a wall. Even when it is heavily fortified, complicated, decorated and elevated to a symbolic urban focal point it is still merely a hole. It is a necessity, but like all necessities a mixed blessing because it allows ingress as well as egress. The fewer such necessities the better; this was always the primeval standpoint and here at Jawa it resulted in the introversion and claustrophobia that seemed to be part of the original plan.

In formalized design such a simple matter evolved along a continuous line that obviously began with natural fortifications such as the cave. Necessarily there are two ways into a cave: straight in or along the cliff; direct or indirect access. In warfare these are defensively and offensively significant directions of approach and we have seen both at Jawa. Yet the cave, far from being man's castle in the architectural, military sense, is his home and it is in the history and development of domestic architecture that the source of the military lies; just as the elements of civilization – the city – are endemic in the village. It is a matter of scale and some additional complication. A gate is a door and in its domestic form all the various tactical elements can be found.

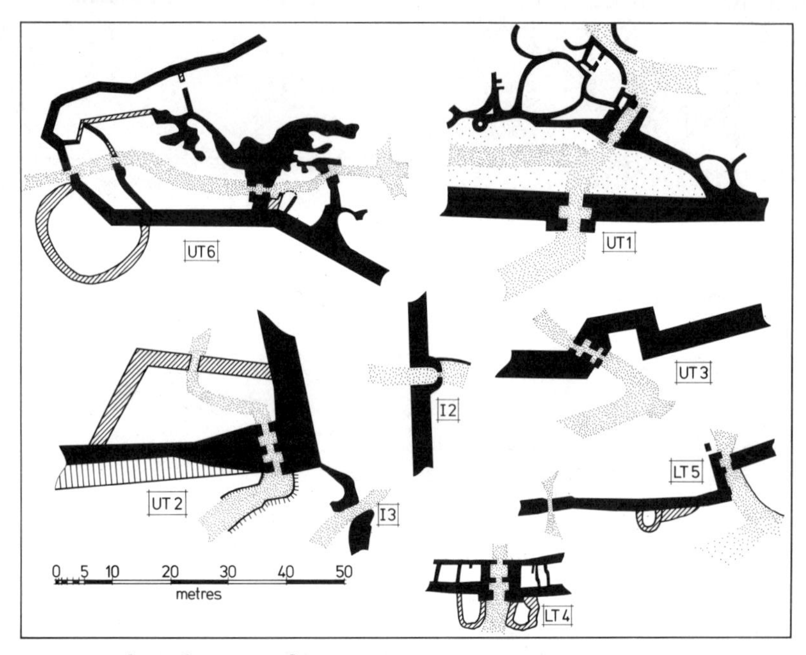

Fig. 45 Plan: the gates of Jawa

The cave was natural form; the house of Mesolithic and Neolithic man of his own making and hence architecture. The earliest Egyptian hieroglyphs express houses and courtyards in the two basic ways according to direction of entry: direct access and indirect access , just as one can adapt these symbols in terms of primeval gates by adding another door. Thus a simple hut could be depicted as either (direct) or (indirect). The elements and even the form of military architecture, as we describe it in the Near East, are inherent in the simple domestic house. In this

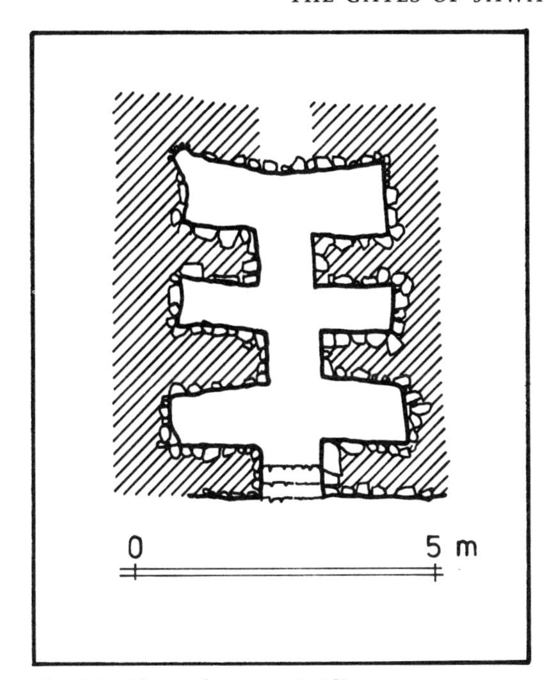

Fig. 46 Plan: a house at Beidha (seventh millennium)

manner one might, for example, take a typical Neolithic house found at Beidha in southern Jordan and add a second door. The Beidha houses have one door and a series of internal buttresses intended to carry roof beams. The new door opposite the entrance would make the structure into a triple-chambered, direct-access gate just like gate LT4 at Jawa – and, but for scale, like much later gates (figure 52).

The humble origin of the mighty gate can also be seen in the Chalcolithic period of Palestine and the broad single-roomed houses with benches that were to become the generic dwelling places during the first urban stages there. At Ein Gedi near Masada on the west shore of the Dead Sea such a gate is actually built into an enclosing wall and is definitely a close parallel for the single-chambered gates of Jawa. Ein Gedi also gives us the simpler form of such gates consisting of slightly reinforced jambs. Its equivalent at Jawa is the simple postern. The whole of the Dead Sea complex – called a shrine by the excavators – symbolizes compactly the embryonic city-in-the-village idea advanced by Lewis Mumford

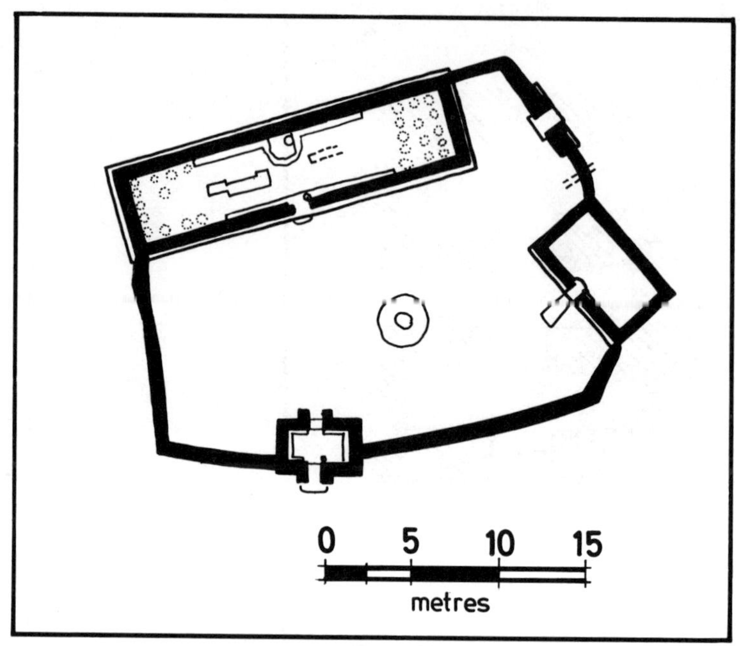

Fig. 47 Plan : Ein Gedi 'Sanctuary' (fourth millennium)

(1961): fortifications, shrines, wells and so on. In the military sense the site includes details such as projecting building lines that are the prototype for wall towers in more formal urban architecture.

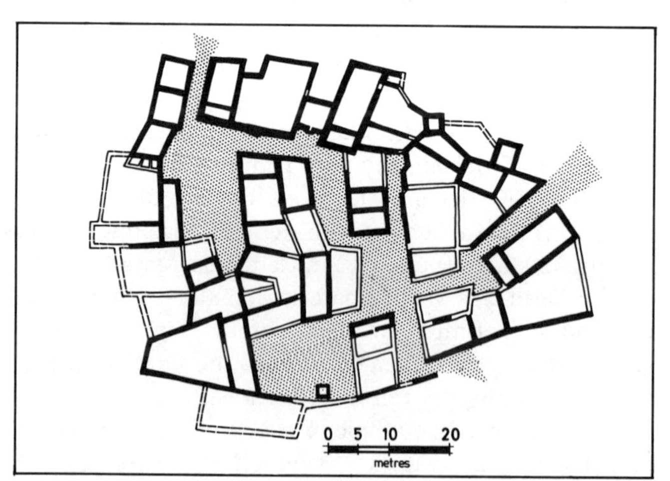

Fig. 48 Plan : Tuleilat Ghassul IV (fourth millennium)

Similarly, the site of Tuleilat Ghassul to the north of the Dead Sea, also dating from the same period as Ein Gedi, can be significantly reorganized in terms of its later stages. If this is done the domestic and village origin of urban fortifications is represented by the joggling projections of adjoining houses. These are the ancestors of the intended trace: individual houses becoming towers, small openings in fences becoming posterns and larger ones flanked by projecting houses turning into main gates with flanking towers. These main gates are of course very similar to those at Jawa – in principle – and only the internal buttresses are missing.

This design, at Ghassul still a primitive formation stage, blossomed in southern Turkey close to this time into a deliberate military architecture. At the site of Mersin (period XVI) the outer house walls have become thicker. They are now proper curtains with slit windows and the gateway is flanked by two specialized

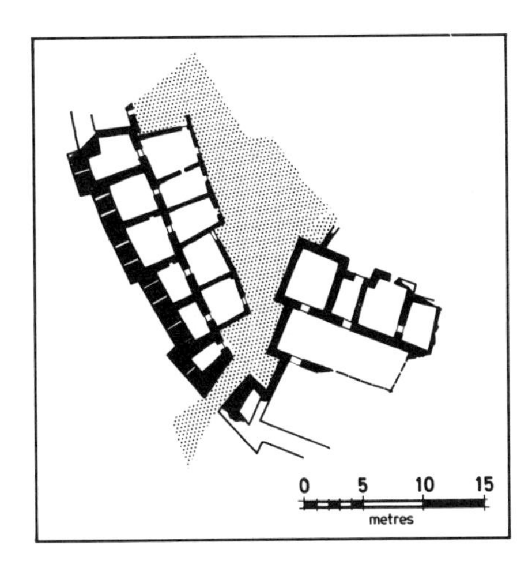

Fig. 49 Plan: Mersin XVL (fifth millennium)

towers. Also in Turkey there is a direct parallel for Ghassul at the site of Hacilar where we can recognize prototypes for internal buttresses as well as both types of gate: the direct-access one, with flanking internal domestic units that function like towers, and the indirect.

The Jawa repertoire of gates is the best collection preserved, all in one unified military urban scheme, and represents a more formal stage in the development. But we understand that the degree of formalization into schemes does not necessarily develop along a

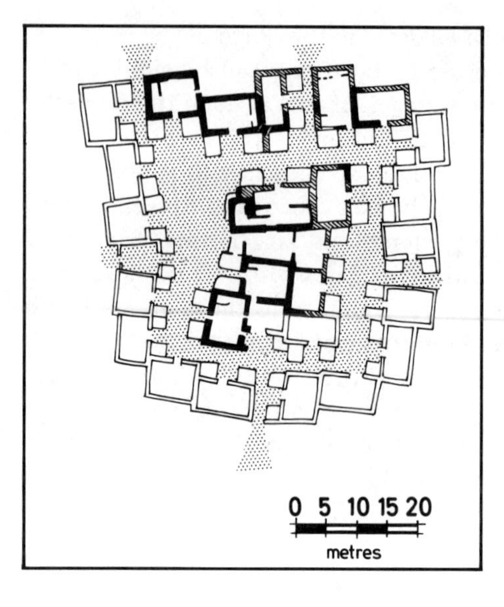

Fig. 50 Plan: Hacilar (fifth millennium)

continuous line: only the elements. This was pointed out above with Tell es-Sawwan where a similar high level of formal development was achieved long before the examples cited here. We saw there a striking similarity with military architecture of the third millennium. The Aegean site of Kastri (figure 33), interestingly enough – continuing the theme of domestic origins for the military – is also an example of the single-chambered, direct-access gate derived from domestic architecture. This gate is made out of a house adjoining the curtain. Note also – as at Mersin – that such 'cells' along the inner side of a curtain in effect make the wall into a casemate: just as at Jawa about gate LT4. Gate UT1 finds its specifically Palestinian cousin at Early Bronze Age Tell el-Far'ah (near Nablus); even Jawa's

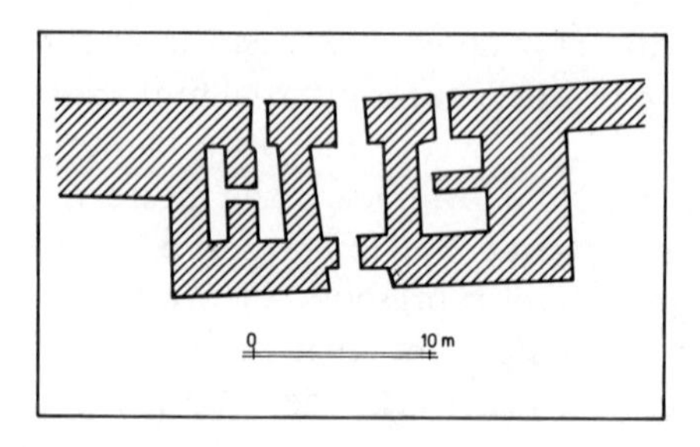

Fig. 51 Plan: Tell el-Far'ah (third millennium)

gate LT4 is related here because the passage between the flanking towers at the later site slopes up and was provided with wooden steps.

Finally, the design of gate UT3 and perhaps also UT2 and the composite of gate UT1 and the ramp can be traced up to and beyond the Iron Age of the Near East. The fully developed scheme shown here comes from Gezer in Israel over two millennia later.

> And there were three side rooms
> on either side of the east gate;
> the three were of the same size.
>
> (Ezekiel 40:10)

Fig. 52 Plan: Gezer (Iron Age)

13

STRATEGY AND TACTICS

From the strategic point of view Jawa and its fortifications are not very advanced.

Returning for a moment to the ideal concept of the town at the beginning of settled life at Jawa, there are severe limitations regarding security and defensibility in spite of the admittedly advanced design and location of the gates. The fortifications of Jawa are an example of the worst kind of static defence from the very beginning, made worse through the various stages of compromise up to the end of phase 2. Although the choice was limited so far as site location was concerned, certain negative factors in the design of the defences tended to aggravate this essentially static, introverted fortification. Jawa was frozen in tactical as well as strategic terms primarily because it had no towers, apart from the elaborate gates, and because it had no posterns. There were, ironically, too few holes in the wall and accordingly no flanking manoeuvres could be carried out with ease and speed. In case of attack on the upper defences, only forward fire could be brought to bear and the strategy of defence in the early stages of battle would be limited to anticipating the point of assault. If anticipation matched actuality, all the defenders could do was pour fire upon an enemy who could advance under cover of the densely built-up lower quarters.

It is true that gates such as UT6, UT1, UT(7), UT4 and UT3 could give lateral fire along their adjacent curtains, but this would be very feeble since only one or two archers or slingers could shoot at the same time. The absence of towers therefore could be taken as a sign of ignorance concerning the tactical importance of the ground immediately before the perimeter. The Jawaites were probably not really aware – until it was too late – of the need to keep this area below their massive upper-town fortifications clear of obstructions, although they did keep a ring-road within. However, even if they had planned such a *cordon sanitaire* or 'security belt' it would not last long. The tactical moment (Yadin 1963) would approach very quickly in any assault on the upper town and enemy troops would reach the crest of the long straight walls. Defence and control of Fortress Jawa would then depend on hand-to-hand combat.

The upper town was isolated on an island with steep sides. The wadis around it may have acted as a natural moat, but the ground outside Fortress Jawa was not controlled: it was allowed to be built over with a rabbit warren of huts and winding lanes that cut off the water supply as well as easy approach to the outer and first line of defence. Lower-town structures also cut off a quick route of retreat. Any conflict that took place would be an urban war, hand-to-hand, house-to-house. No broad strategy was really possible because of the obvious lack of rapid communication possibilities, easy movement of troops, clear field of fire, effective artillery and so forth. Furthermore, Jawa was cut off from its own food supply although this may not have been too serious since ample storage facilities lay within the upper town. This, however, was less a matter of military planning than traditional domestic architecture.

It must be accepted that Fortress Jawa as a concept was not a success, which is a little hard to understand in view of engineering brilliance elsewhere at the site. One explanation of this strange discrepancy might be that the Jawaites were simply too confident in their own might and urban superiority over the aborigines; another that an immediately hostile atmosphere left them little time to build anything more sophisticated than a stone fence, although they knew better. Perhaps they were specialists only in hydro-technology and not warfare. Perhaps that is why they lost their urban roots in the first place.

It has been seen how the upper town was eventually cut off on all sides through a series of planning compromises so that the only preventive measure along military lines lay in the elaboration of the gates. However, an attempt was made to control the situation, to correct the weaknesses of the system and to retain open access to the water supply. In these factors should be recognized what is often suspected but rarely seen in prehistoric cultural development. During this phase – need being not so much the mother of invention as of adaptation – Jawa appears to exemplify the hard process of learning how to change and compromise: to make intelligent concessions to reality. The changes that we suspect took place in the minds of the Jawaites did not generate a new system during the short time that was to be Jawa's urbanism in the Black Desert. Perhaps in failure and ultimate disaster a positive result or even a new consciousness took form later in another, kinder place, reversing for once the trend of not learning from mistakes. These ideas only germinated at Jawa; but the seeds may have been transplanted, as had been Jawa's technology up to this point, and may have grown into a new concept of urban military architecture in Palestine to the west.

But for the time being the science of war was still essentially non-urban at Jawa. It was still uncivilized, and in that sense Jawa stood between the village and the city.

This median position is supported by Jawa's arsenal, such as it is: exemplified by being unspecialized. The tools of the hunter, the farmer and the shepherd were made into weapons, just as the symbols of royal proto-imperial levies on Egyptian pre-dynastic palettes are shown in war with adzes and hoes while the humble draft animal is taken from behind the plough to become the first mechanized battering-ram.

The flint implements of Jawa (Appendix B) give us the bow and arrow as well as the medium-range spear. The abundant stone everywhere provided ammunition for the sling and the two arms, which together constituted the artillery of the town whose full effectiveness could not be realized within its defensive system. Pierced and drilled stones – primarily the hoe and adze – became the club and primitive battle-axe. Only the polished stone mace-heads represent a specialized symbolic as well as tactical weapon in which one may see the officer's side-arm or the sceptre of a king, just as the compound-bow became the special weapon of its day at the end of the third millennium BC.

However unsophisticated the weapons – a state of affairs that prevailed throughout the next millennium as well – the potential combatants at Jawa were not an undisciplined rabble. There was a central guiding intelligence that controlled and directed the people of the upper town and those of the lower quarters. This was the same intelligence that fought the losing battle of planning and the same collective brain that designed, built and maintained the brilliant water systems, ironically and perhaps fatally understanding nature better than man. Discipline in any field, once acquired, can be turned to war. For the Jawaites this discipline derived from the organizational craft needed during their recent exodus and that developed in their urban roots; for the others at Jawa it came from their tribal bonds and experience of nomadic warfare – raids and blood feuds but few pitched battles. The Jawaites would be structured into some kind of citizen militia while the sub-urbanites would probably have served more as mercenaries in times of siege, responsible for the defence of their own sector in the lower town under the direction of Jawaite officers.

The kind of conflict that could and would have occurred at Jawa hardly seems worth the name of warfare which implies large numbers and scientific manslaughter. On the other hand, it would not have been a completely disordered brawl – at least not at the start of hostilities. There would have been the potential for two types of war: a war of attrition with the desert folk at large leading to a siege of Jawa, or a civil urban war with all of the traditional trappings of riots, mobs and barricades in the narrow lanes. The action would entail limited use of projectile fire and a large amount of hand-to-hand combat. It would rapidly become the miserable, messy but nevertheless deadly *mêlée*, no matter how well planned, that would ensue once the gates had been broken through and the walls scaled or breached. Then, depending on how the struggle went, there would be pursuit into the desert or retreat into the town until the last defender was clubbed to death. The usual burning, rape and pillage would follow. Once more we are drawn to the edge of the abyss.

14

DOMESTIC ARCHITECTURE

As it was for the gates of Jawa, the domestic architecture represents a stage in a long development from man's earliest shelters. Along this evolutionary route specialized forms occur in accordance with the mechanics of such processes: adaptation to environment and, later, man's whim – style. But as the gates were simply holes in a wall, so these Jawaite houses were just simple walls and roofs: shelter first of all from the weather and then from fellow man.

The two basic architectural forms, the circle and the square, appeared virtually at once. From nature and the cave came the round hut and from man the rectilinear. Both forms were there from the beginning because leather, stone and earth give shape to primordial architecture in which the building block or module does not necessarily determine shape. Stone houses can be round as well as square whether the stone is worked or not. The brick, whether made by hand or in a form, produces both shapes in plan and elevation (arch, vault or dome). Throughout the earlier development of the domestic house and up to the time of Jawa stone and form-made bricks were used in both ways and so it continued into the next millennium, the Early Bronze Age. However, while the module and hence the method of assembly do not preclude form, they do naturally affect it: if only for the simple reason that the round is less stable than the square. The latter therefore tended to become the dominant and the rectilinear the predominant form of architecture until today despite new materials and advanced technology. A geodesic dome is still made up of planes.

Predominantly rounded (natural) stones were used in the foundations at Jawa and determined the plan. Thus most dwellings have more curves than sharp corners. Interestingly, when it came to the upper brickwork the Jawaites had to face the problem of making a curve with rectilinear form-made bricks. The rounded form in the history of architecture becomes less normal and notably is employed for special purposes: in towers, graves and public buildings, both secular and religious.

We know that the basic dwelling unit at Jawa is unspecialized

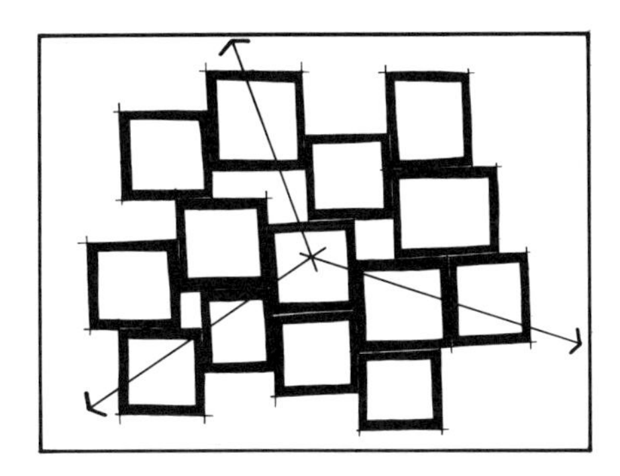

Fig. 53 Schema: Cellular

throughout the site. It was simply an internal living-sleeping-working-eating shelter, an external working area and one or more storage bins. There are also examples of extended family aggregates, hence unit growth by repetition and addition. The space between

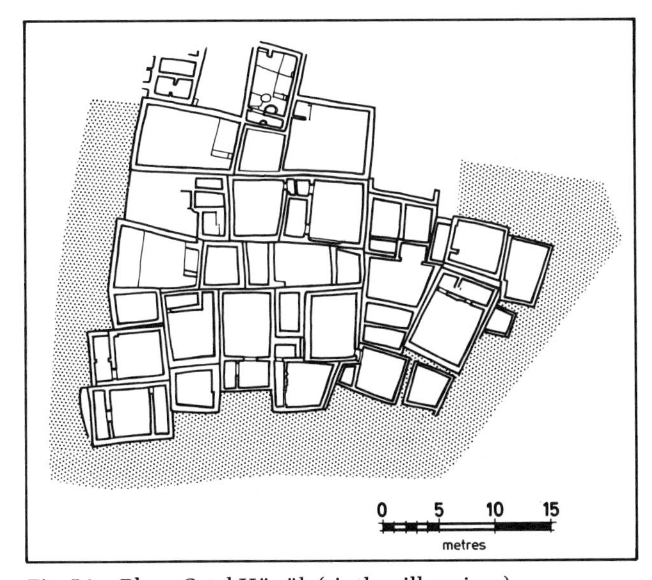

Fig. 54 Plan: Çatal Hüyük (sixth millennium)

these super-units determined the nature of the settlement as it developed along predictable lines: the units grew in a finite container (the fortifications), outwards but towards each other.

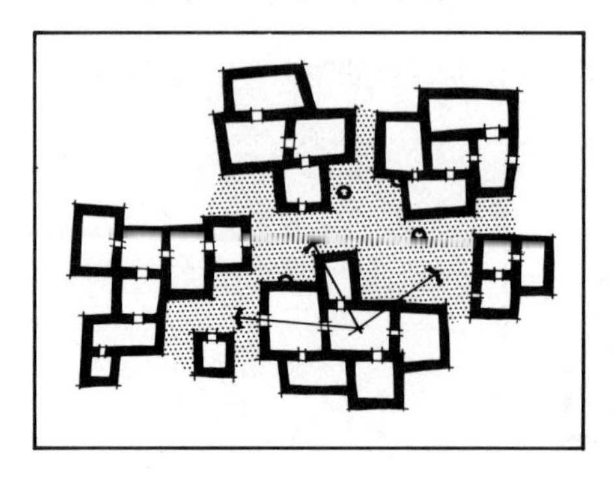

Fig. 55 Schema: agglutinating insulae

The units grew by addition, not by division like a cell but rather like budding yeast, from the core out with the added part eating up space. Settlements known from archaeological discovery throughout the world follow roughly two fundamental patterns of such agglutination: a centre to periphery free expansion and a

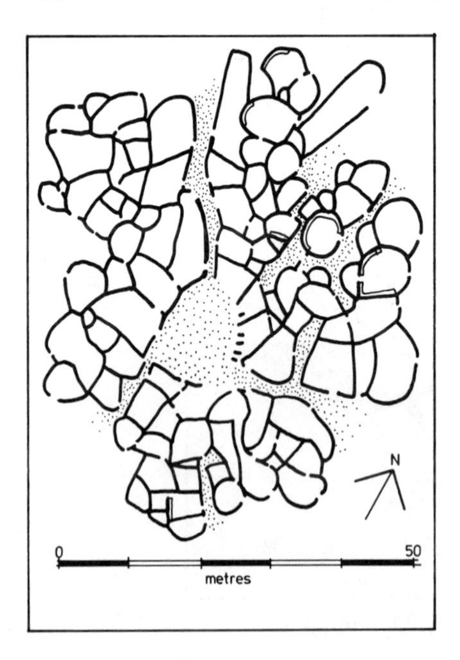

Fig. 56 Plan site 688 in Sinai (fourth millennium)

densification. The first is well illustrated at Çatal Hüyük and precludes a separate defensive wall since the living units make a blank wall-face to the outside world. This form finds its modern equivalent in concrete bunkers or geometric hive-like domestic complexes. The second pattern of growth has been called the phenomenon of agglutinating insulae (Schmidt 1963; Heinrich 1964). This has already been observed in the development of gates.

At sites like Hacilar, Tuleilat Ghassul, even Ein Gedi and in Sinai site 688, units grow together to form proto-streets between them and an external barrier either by joining house walls or separate houses with yard walls. At Mersin the curtains, though thickened, are still organically part of the domestic architecture. At Tell es-Sawwan long before this the separate curtain was the container. The pattern is well represented during the third millennium at Kastri, but it should be noted that within it any amount of formalization is possible, although perhaps not as a rule. An example is the apparently advanced form of Tell es-Sawwan, while many Neolithic villages possess remarkably regular streets that are normally considered as a later stage of the agglutinating insulae development. This 'anomaly', among other criteria, led Mellaart (1975) to call them all cities.

Jawa belongs, as before, to a median position within this pattern: to a stage that is illustrated in contemporary Egyptian palettes. The space between expanding domestic units is still communal. Such space becomes the street network that in the broader history of the city develops towards the increasingly impersonal, just as private space becomes more exclusive and public activity is channelled and concentrated in special structures and areas. The streets at that stage are only communal in terms of traffic.

The houses of Jawa are all alike in form and structure. Many were partly subterranean where the ground surface permitted this and all of them had beaten mud floors, stone foundations, plastered walls, mud-brick and plaster superstructures, wooden roof supports on stone base plates and roofs made of wood beams, smaller slats, reeds and mud. Doors – where we have found them – were few and when a house was let into the ground the doorways had a stone step beyond the threshold flanked by a door socket. Some houses have stone benches along some walls; some have internal divisions. All were fitted with small pits, hearths and at least one rounded stone-lined storage bin reached from the living area through a low door. These bins were probably domed. There is no evidence for windows as the

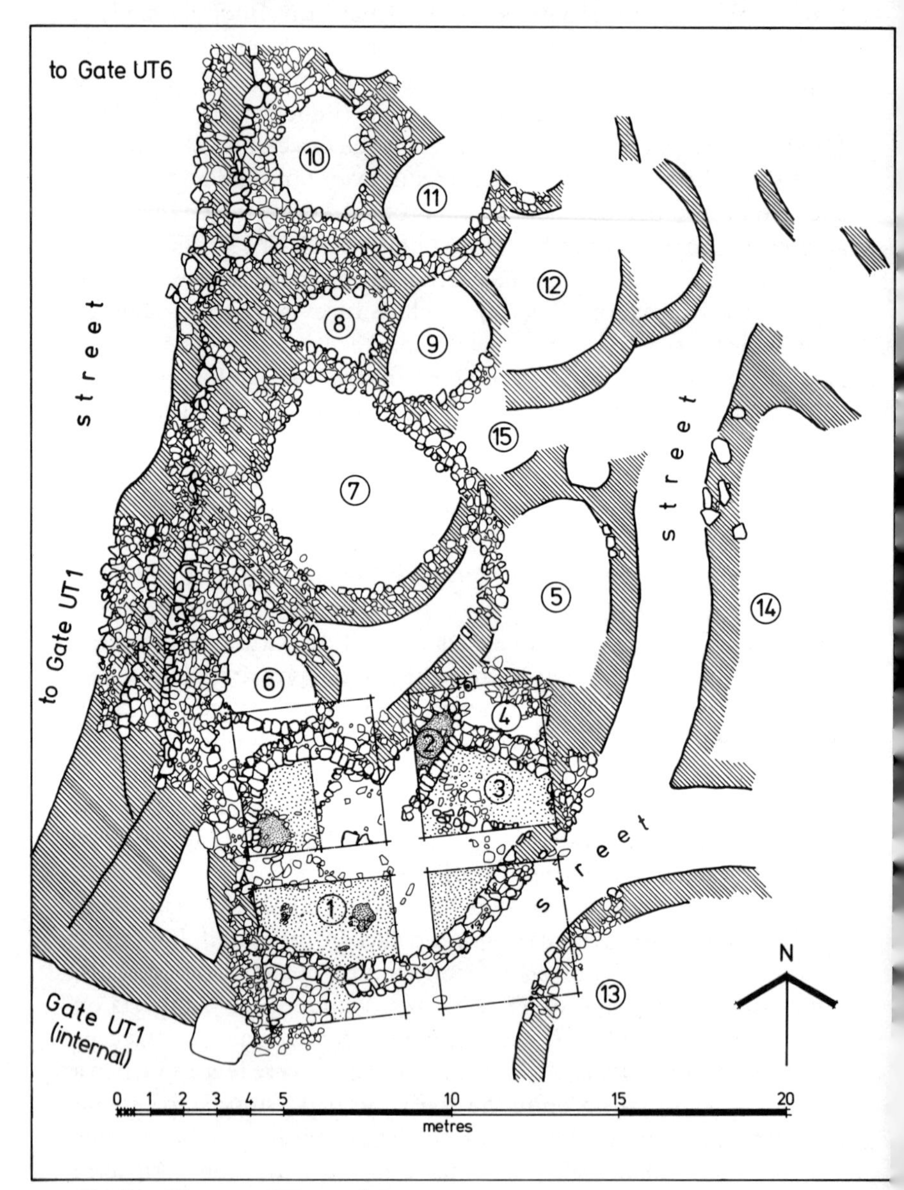

Fig. 57 Plan area UT trench I

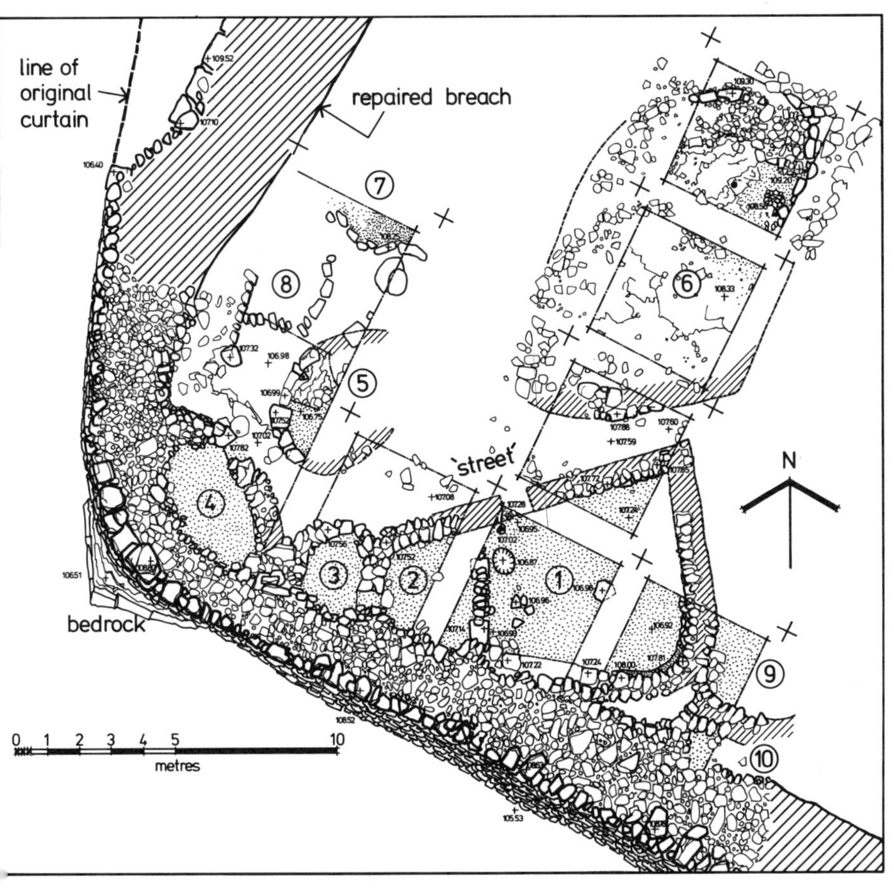

Fig. 58 Plan: area F trenches II–IV

walls stand today only a few courses above the ground, but it can be assumed that small slits up under the roof would let light in and cooking fumes out. A typical domestic unit is shown here. It belongs to phase 3 but represents all stages at Jawa.

Architectural parallels abound and some are topical concerning a development of a very general kind. It may or may not be significant that the houses at Sinai site 688 have benches, or that certain Chalcolithic dwellings at Beersheba in Israel appear to develop from totally troglodytic through semi-subterranean to surface structures. It might be relevant that the shrine at Ein Gedi, also Chalcolithic, has these benches, especially the long room there which is exactly the

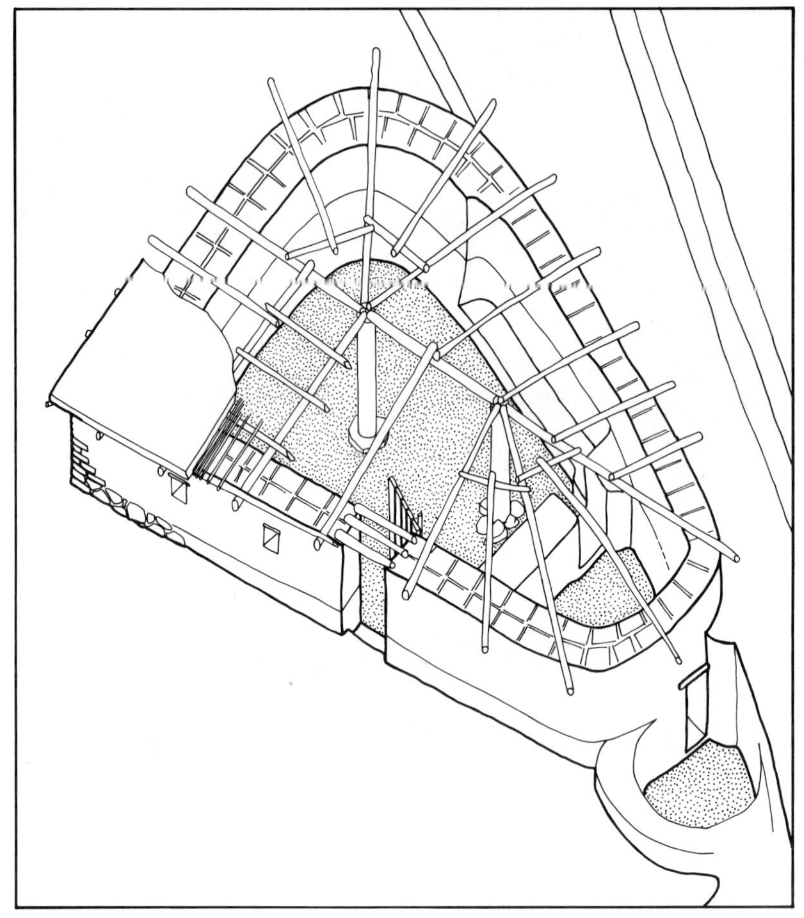

Fig. 59 Reconstruction of a house in area F (phase 3)

prototype for the standard Early Bronze Age house a few centuries later. Still, by that time the typical dwellings at Tell Arad for example are also let into the ground and entrances all have steps down and door sockets on one side. Just as relevant are the physical limitations imposed by necessity, common usage, common raw materials, building methods and the natural environment. What else could one make from mud and stones? The environment may be related in a more general way by comparing the semi-subterranean nature of some Jawa houses to a quite modern practice in the desert.

While we were working at Jawa we often visited outposts of the Jordanian Army and saw their tents, which had been dug into the ground. They would excavate a neat rectangular hole and throw the soil up around it as a low wall. The tent acted merely as a roof while the earth bank, sometimes reinforced with stones, and the low profile of the finished structure protected them from the incessant desert winds.

Only one house deviated from the norm. It was found in trench UT IV and had two right-angled square corners in one wall. Like the rest it was provided with internal divisions and surrounded by stone bins. The deviant geometry of this small house cannot however be taken as a deliberate specialization (it is not a shrine) and it has to be concluded that sociologically – inasmuch as this relationship may be valid – the very undifferentiated domestic architecture of Jawa might point to a rather egalitarian society and that only the difference in fortifications between the upper and lower towns hint at something else, as has already been argued. The apparent lack of large and dominant religious architecture may mean that the people of Jawa practised a kind of pragmatic naturalism (water worship?); or, more simply, at subsistence level could not afford the time and effort for such expression in physical manifestation.

The only truly specialized installations we found come from the

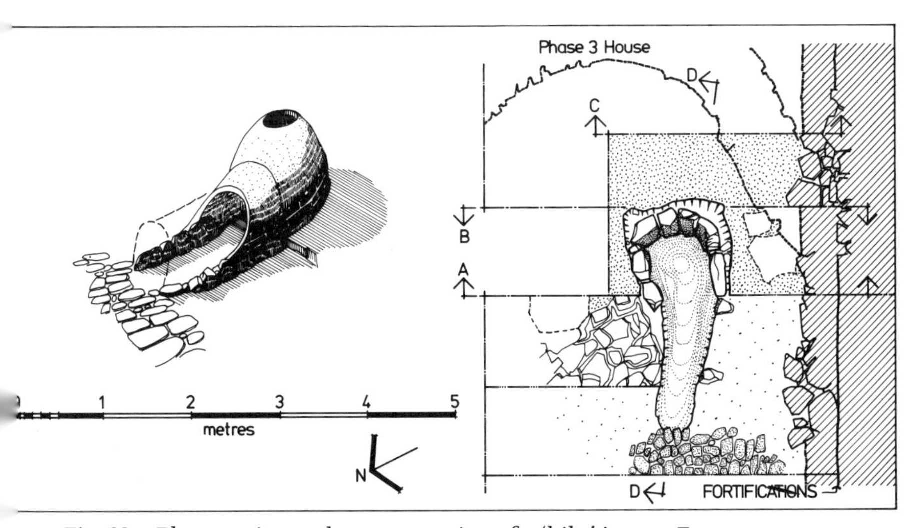

Fig. 60 Plan, section and reconstruction of a 'kiln' in area F

upper town in area F and belong to phase 2. The best of these was difficult to identify in the field and some 'copper' slag found nearby led us to think in terms of smelting or re-casting furnaces. With the slag was some of the original fuel which, since we had been walking on its modern equivalent for some time, was more or less immediately identified for what it was. It consisted of sheep or goat dung. The slag and other information, however, soon provided a more realistic answer: 'The slag is certainly vitrified fuel ash. It could have come from a pottery kiln or something more domestic (a bread oven). It has certainly been fired to a high temperature by a forced or well-induced draught' (Tylecote in Helms 1976).

This and one other installation may therefore have simply been domestic and part of a nearby house. On the other hand, if they were pottery kilns we might be able to hint at industry and possibly its specialized location within the town. One very obvious factor limiting this interpretation is the need for draught which would be difficult in area F because of the nearby fortifications: unless of course there was a gate under the later repaired breach where the position of a gate, UT(8), is postulated. Another point to be noted is that an industrial area would hardly fit into the scheme of a patrician upper town unless, as is possible, the craft of the potter was exclusive to the Jawaites.

15

CENSUS

Since so very much is visible on the surface and since it was possible to record most of it from the air in relation to excavated areas, it is now possible to perform some quite realistic calculations regarding the population of ancient Jawa and to justify the figures used in the earlier hypotheses in this story.

The aerial photographs of course represent only the final stages of settlement (phase 3) and not necessarily that of the 'classical' stage at Jawa. But as has already been seen, there is sufficient evidence – architectural as well as in terms of planning and layout of the town and its fortifications – for reconstructing this stage. Combining excavation results and aerial survey, it is found that the average dwelling unit has an area of about 50 m^2. In the test areas that we chose – and it is noteworthy that the unit size and other relationships are virtually the same in both upper and lower towns – the following average spatial relationship to elements within is as follows:

Single dwelling units	46 %
Storage bins	10 %
Yards	22 %
Streets or open areas	22 %

When applied to our standard test area within the domestic sectors this gives the average number of single dwelling units per area as 4·3. Taking the various population figures in relation to housing as well as urban and village density throughout the Near East and also results from the research in the Negeb Desert (Evenari 1971) we can assume that each unit housed between four and six people, giving an average population range of 17·2–25·8 per test area (470 m^2).

Before this can be applied to the site of Jawa some adjustments must be made regarding the total external area of the town, which is about 120,000 m^2. Also the two ethnical sectors of Jawa should be treated separately.

The total area of the upper town fortifications is about 4680 m^2, to which must be added the area of the gates (*c.* 3350 m^2) and the area of the ring-road (*c.* 3500 m^2) making a deduction of about 11,530 m^2 from the external area of the upper town (*c.* 53,100 m^2), leaving about 41,570 m^2 for our calculations. This represents 88·4 test areas and a population range between 1520 and 2280 people. For the lower town the total area of the defences is in the region of 4648 m^2 and the circle roads would add another 12,000 m^2 or so, requiring a deduction from the external area of the three layer quarters of about 16,648 m^2. The lower town figures are as follows:

West quarter		18,200 m^2
South quarter		18,850 m^2
East quarter	(a)	15,900 m^2
	(b)	2475 m^2
	(c)	11,975 m^2
		67,400 m^2
less		16,648 m^2
		50,752 m^2

Thus we have 108 test areas for the lower town and a population of between 1858 and 2786 people giving Jawa a total range between of 3378 and 5066. This represents 18 m^2–17 m^2 per person, 0·05–0·06 persons per square metre, or 148–222 people per acre; a low range when compared with population densities in some modern settlements that are occasionally taken as a guide for early towns. The figures should also be understood as an ideal: because the population density of Jawa was planned and (hopefully) controlled. Any growth may have been absorbed for a while by the expansion of the lower quarters. As was seen earlier however, this could not go on for ever simply because of topographical limitations of the site.

In general and as far as the story of Jawa has so far been taken, one observes a substantial achievement and a high level in civil planning, the execution of these plans, their change, adaptation, compromise and control. In purely technical terms, there is the probability of advanced building techniques: the corbel and the relieving triangle, the form-made mud brick, a knowledge of surveying (the straight long walls, for example), an evolving if slightly primitive military architecture and, most important of all, the concept of a planned ideal town with the understanding of compromise. Thus Jawa demonstrates all the problems of the city at a very early stage in its history. It also reveals the ogre of endemic obsolescence.

V

LIFE-SUPPORT SYSTEMS

Detail from a predynastic Egyptian mace-head (Ashmolean Museum): a fourth-millennium king ceremonially opening an irrigation canal

16

WATER RESOURCES

The structure of the Black Desert precludes access to ground-water and springs are very rare indeed. It will be seen that only water derived from the brief winter rains could transform this bleak land momentarily into a life-supporting environment and that without highly organized and developed technology that moment would pass and the land revert once again to the sterile lunar appearance that greets the summer visitor.

That water should occur naturally at Jawa and not pour from the back of a tanker truck was hard to believe as we worked there during the dry season. I had been at Jawa in the winter and witnessed the abrupt, violent abundance of water crashing through the wadis, across roads, washing out bridges, flooding kilometres of the desert sometimes within a few hours of the first rain drops; I had even seen hailstones bringing down birds in flight. Nevertheless, especially when one is far removed from the harsh reality of the place, what is most clear in memory is the unrelenting dryness, an absolute lack of natural water and thus a reluctance to concede that anyone human could ever have lived there for long. How could a large permanent settlement have survived? Yet the ruins – lanes, streets, houses and fortifications – are there as irrefutable evidence.

There are also the remains of Jawa's ancient water supply which seemed at first glance not very easily comprehensible, although the big dam west of the site is unmistakable in its intended function. It was this structure along with the ruins of the town above that were particularly noted during Winnett and Harding's epigraphical survey over twenty years ago. Today as one walks about Jawa a number of features begin to emerge that seem to be related to water, quite apart from the work of man, belying the deadness all around.

In the big wadi east of the site there are large basalt boulders wedged in the tops of hawthorn bushes,some of the stones hovering nearly 2 m above the stream bed. Nearby other signs of such 'environmental' sculpture are evident. Wide expanses of strangely smooth basalt curve gently in the sunshine, almost alive, and along the steeper cliffs there are sharply scalloped ridges, hollows and hyperbolic shapes and voids like the work of a surrealist. On the flat basalt of the wadi bed one encounters perfectly rounded parabolic depressions that seem to have been carved mechanically; and always, at the bottom of such holes, one finds a small pebble worn round and shiny, like a tool only recently laid down.

The sculptor of these works is and was of course wadi Rajil, the life-source of Jawa, and the winter floods that have come rushing down each year from Jebel Druze ever since the lava cooled continue to carve the black rock of Jawa into an ever-deepening canyon.

Still other signs indicate that Jawa was not always dry. Next to the ancient town, in fact within one of the old reservoirs, some bedouin had excavated a wide, deep depression with bulldozers, pushing the yellow silty soil up in a great crescent-shaped dyke. This was done very recently since nothing of the sort was there in 1966. When we began digging up Jawa in September 1973 there was still a little moisture at the bottom of this modern pool and around it in the drying mud the prints of every desert animal known in the region: from camels to tiny song birds, from snakes to insects, all converging on the receding circle of drinking water. Here was the latest testimony to Jawa's hydro-technology, at the very least attesting to the correct choice of site since the depression still collected run-off from the winter rains that fall on the rolling hills to the west, just as it had done during the first phase of Jawa long ago. There are even signs that ancient canals and other parts of the water systems were re-used quite recently and some of our bedouin workers actually understood these things better than we did at the beginning. But what is most apposite to the Jawa story, and was a little frightening at the time, was an event during the 1974 season of excavations.

Early one morning – we had been getting on with our day as usual – a truck whose whine we had heard for over half an hour, suddenly appeared on the ridge west of Jawa. Since we were used to being almost completely alone except for occasional visits from the Jordanian Army (whose trucks we know) all work stopped. A dozen or so heavily armed bedouin dismounted and gathered in a tight and sinister knot, ignoring us and our camp entirely and far more interested in the ruins of Jawa. So we all stood, a sort of frozen pantomime, until the noise of a second truck, this time from the opposite direction, from the north-west and Deir el-Kinn, made us turn that way. Brakes squealed and there on the rocky spur overlooking Jawa from the north-west stood a second silent group of armed men, also watching the site and the others – but not us. We seemed not to be there at all, which was no relief as more time passed and some of us began seriously to look for cover and wonder how much a heavy army tent might slow down a bullet. Then yet another vehicle was heard approaching, this time with a high-pitched sound, faster and more urgent, and around the corner came the police and army, bumping into the camp in a streaming command-jeep to join what seemed like a silent tribute to the black ruins of Jawa.

As if by signal everyone moved – everyone except us. We just stood there as if we had all looked back at Gomorrah. The two armed bundles on opposing hills spilled down the cliffs and moved towards each other as the army and police split in two to head them off. The whole drama came to an end at the centre of the new yellow pool on the dried footprints. There was a lot of talk, many gestures, some fairly quiet words we could not understand over the distance, and then everyone left Jawa.

Our rescuers stayed a while to drink tea and told us what it all meant. One family (south-east truck) had hired a bulldozer to deepen the ancient pool so that it would collect winter rain to be used by their flocks and camel herds. The other (Deir el-Kinn truck) had taken water for their animals. That was considered unfair and it was felt that some money should change hands. There had, they said, been a little shooting before we came to Jawa that year. It was referred to as a water war, which is why our rescuers were so quick to arrive. How they knew of the confrontation is hard to tell, unless they too heard the truck while it was still kilometres away. That apart, as we returned to our work amid the agitated discussions of our workers – back to the fourth millennium and peace – the big yellow pool by the side of Jawa took on a deeper meaning.

While working at Jawa our daily water was of course supplied by Jordanian Army tanker trucks. This method of water supply is the most common one used today and relies on water pumped and piped from the oasis of el-Azraq to the Mafraq road where there is a free tap. All may take water which is moved away by the bedouin according to their wealth: by truck or jeep, camel, donkey or on foot. The water is stored in jerry-cans, old oil-drums, rubber – no longer skin – bags, tin cans and also often in ancient cisterns and pools originally built by the Nabateans, Romans and others who seem to have understood this desert better in those days. Unconscious reference to even more ancient hydro-technology may be seen at Deir el-Kinn where the now semi-settled bedouin scoop out shallow pools, heaping the spoil in half-moon ramparts like the contested pool at Jawa. The open end points uphill and the pool is filled by surface water (run-off) that is frequently assisted and directed by artificial gullies and even low stone channels radiating out across the gently-sloping land. In much the same way water is collected in rockier, hillier areas to the south where we have found shallow depressions blasted into the bedrock with dynamite. The loose rubble is piled around the 'pool' and stone-lined gullies, sometimes only one stone high, fan out as at Deir el-Kinn across the land much like the long arms of the desert 'kites' discussed earlier. Just like the guiding arms of the 'kites', the lined gullies of the rock pools betray an experimental method of surveying in which several lines are set out to find the most efficient position for catching overland flow. These systems are used by shepherds to augment their water supply for the flocks and to stretch it, if only slightly, into the dry season. These and the pools at Deir el-Kinn are simple forms of surface water collection and storage whose antecendent in principle is ancient Jawa.

Not far from Jawa we examined a third indigenous water collection and storage system the basis of which was a gift of nature stemming from the first formation of the land. This relied on harnessing surface water in a matrix of shallow canals that eventually, after several kilometres, led to a natural lava-flow cave. One of these 'systems' ends in a cave called el-Mughara (cave in Arabic) about 8 km south-east of Jawa. It is a long low-vaulted tunnel now filled with silt nearly to the roof, but still damp even in mid-summer. Water stored here would last much longer into the dry season because of the shelter from the sun which would reduce evaporation losses considerably. There seems little reason to doubt that such systems are ancient, although perhaps re-used in more recent times. The silting is a slow process, and the caves were made as recently as the Quarternary period (before *c.* 10,000 BC). By the fifth and fourth millennium these natural underground reservoirs could have held thousands of gallons of accessible water. Furthermore, we have often found early flint material in the canals leading into them. It is likely therefore that the use of lava-flow caves and the canal systems associated with them date from the time of Jawa and even long before that, perhaps from the time of the 'kites' and other structures. They may represent part of an indigenous water-collection and -storage technology that may not have been as advanced as that of Jawa but was certainly related to it in principle. In the same way one ought to understand the structures at Deir el-Kinn and the dynamited pools. Both are the works of bedouin who may be regarded as the descendants of the Safaitic nomads who in turn lived much like their possible ancestors millennia before that, beyond the time of Jawa, back to the 'Old Men' of Arabia. Thus one can see one more origin of a kind of hydro-technology that developed locally because it had to: it was the only way in which man could exist in this desert.

When leaving Jawa in 1976 and passing by Jebel Rima on our way to the surfaced road for the last time we saw that the settled bedouin there – partners in the water war of 1975 – had built a large circular pool of heavy concrete segments to collect water. The design of the new pool is the same as that of ancient Jawa's reservoirs: ironically – but for the concrete – less sophisticated. In the days of Jawa there were no trucks, no pumps, no pipelines and no water towers like the ones now proposed for the region; but even in this age of developed technology all they have succeeded in doing at Tell Rima is to tap the same water source at two widely-separated points: locally-generated run-off from a very limited micro-catchment and run-off from the wider region (macro-catchment) that reaches el-Azraq along wadi Rajil whence, as was seen, it is pumped and trucked north. The Jawaites did it all at their site. They tapped both sources at the same point. They did it thousands of years ago and supported a settled community numbering thousands in probably the same marginal environment which supports but a few hundred souls today, in small, poor settlements.

Simply because of the environment, because of the nature of the water resources, it is axiomatic that whenever human life existed in the region so did a form of seasonal run-off collection and storage. But one thing is certain: whatever system might have been known before Jawa the size of the new settlement and hence its population density demanded a very much higher level of technology than is evidenced by previous human achievement in the Black Desert. The scope of such systems, moreover, demanded a far more complicated social organization and a developed sense of collective responsibility, a different consciousness also with regard to science, than is found among nomads. The demand for water was concentrated in one place – not dispersed over a wide area as before, when a relatively large population would survive only through mobility in pursuit of game, as well as water. Those people lived more on nature's terms than on their own. To remain in one place throughout the year, and in large numerical concentration, required life-support systems that could challenge nature. In this lies the difference between Jawa and the desert around it: risking the choice of immobility at the centre of an essentially hostile environment. Nature became the servant of man, and with that came the inkling of the idea of man at the centre of his own universe: man as creator of his own environment, urban and civilized man.

Of course the Jawaites seem to have had few alternatives. Their defiance of nature was probably more a necessity of the moment than a conscious philosophy. Yet, as has been seen, the origin of these ideas lay in the development of the city. The idea of man at its centre could – with science and technology – produce a confidence, or a conceit, that man could indeed control the natural world beyond the city walls as he controlled himself within. Without such arrogance it is doubtful whether a primitive animal-survival instinct alone would have sufficed the Jawaites during the first stages of their settlement. It is as doubtful that science alone would have been enough.

17

HYDROLOGY: THE RUN-OFF PROCESS

The hydrological cycle is essentially simple and only becomes complicated when applied empirically, particularly with a view to prediction. To some extent this is what we will attempt to do at ancient Jawa. Theoretically, there is order; empirically, usually

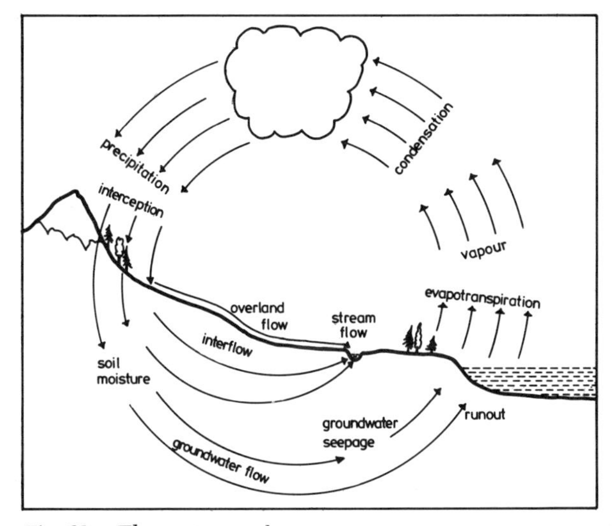

Fig. 61 The water cycle

insufficient data. This is very much the case with Jawa and the hydrological process that governs the life in the region. Much of what follows here and later cannot be measured accurately and is therefore at best hypothetical. However, our task is simplified by the very structure of the land which causes the hard life-style demanded by the Black Desert, notably the thick sheets of basalt and tuffs extruded from Jebel Druze and other volcanoes. Thus a revised or specialized regional water cycle can be shown in which it is obvious that groundwater would never have been a factor of great importance in the water resources at Jawa. A good part of the standard process by which water might reach the site may be eliminated at once, because of the special nature of the region, and surface water alone is of any significance. The manner in which this water moves is the run-off process.

For Jawa the cycle begins over the Mediterranean and Red Seas and occasionally over the Persian Gulf. Condensation of water vapour forms clouds that mass on the periphery of the deserts in winter and sometimes tear across the sky, exploding in violent storms of high intensity and of remarkably short duration. These storms, as far as Jawa is concerned, shed some of their cargo along the way to Jebel Druze, often in isolated local storms that might strike a lucky bedouin's randomly-seeded field. The bulk of water however falls on the mountain. Jebel Druze receives an annual average rainfall of over 500 mm in its upper regions, compared to Jawa's 150 mm. From the mountain the water begins its long

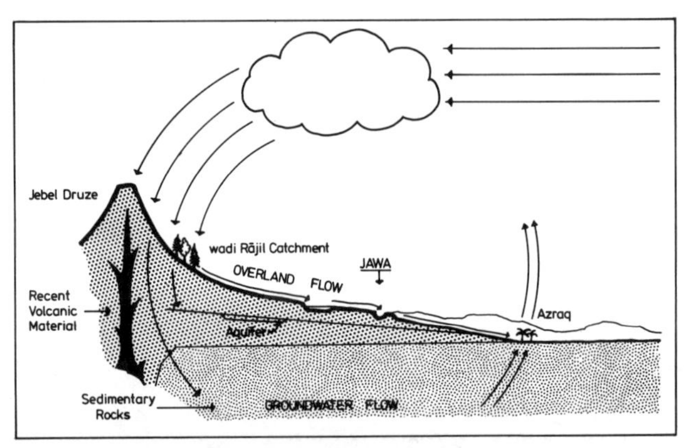

Fig. 62 Water cycle adapted to Jawa region

circuitous route back to lower levels, eventually creating oases and springs along and beyond the edge of the basalt. The maximum ground surface upon which this precipitation falls and that is linked in a drainage system is the catchment: here the wadi Rajil Catchment. It is drained across a series of sloping hillsides, through gullies, rivulets, small channels and finally large wadis of which

1 The second aerial photograph of Jawa (1974) after Poidebard (1931) with the excavation's camp in the foreground

2 Umm el-Jimal: a Roman town made from the black basalts of the desert; the next 'urban' stage after Jawa in these lands (LAH)

3 A typical Black Desert landscape with a small wadi cut into the basalt, debouching into an extensive mudflat. The stone circles represent settlements dating from modern beduin times back to the earliest occupation in the desert (SWH)

4 Qa'a Mejalla area: volcanic landscape in the eastern Qurma Gap, near wadi Ghadaf (JG)

5 Wadi Rajil 2 km south-east of Jawa: an inefficient modern dam cemented against thick basalt and tuff beds; the frozen lava of the Jebel Druze volcanoes

6 Wadi Rajil 2 km north-west of Jawa: a box canyon carved by annual floods, exposing thick layers of basalt which weather into natural 'megalithic' building blocks. One of ancient Jawa's canals passes only 100 m further west (SWH)

7 Qasr Burqu' on the eastern edge of the basalt: a Roman and later Islamic frontier fortress beside a perennial water source

8 'Animal Farm': no longer so enigmatic ancient carvings of cattle (LAH)

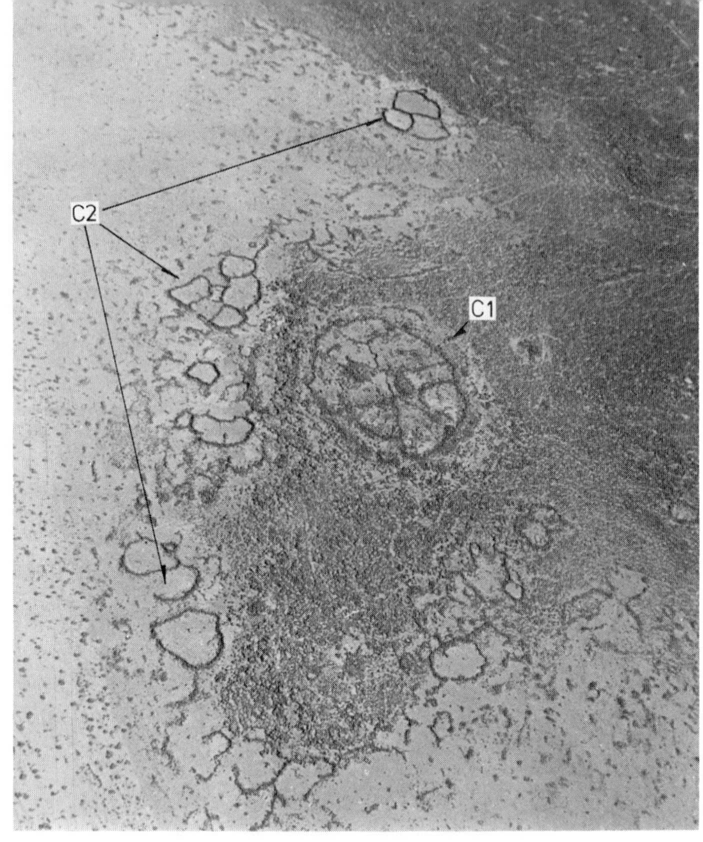

9 Ancient occupation in the Black Desert: the hut-circles described by Maitland and Rees in the 1920s; in the centre a 'jellyfish' (C1) c. 50 m in diameter, surrounded by other sites of various periods (C2) (SWH)

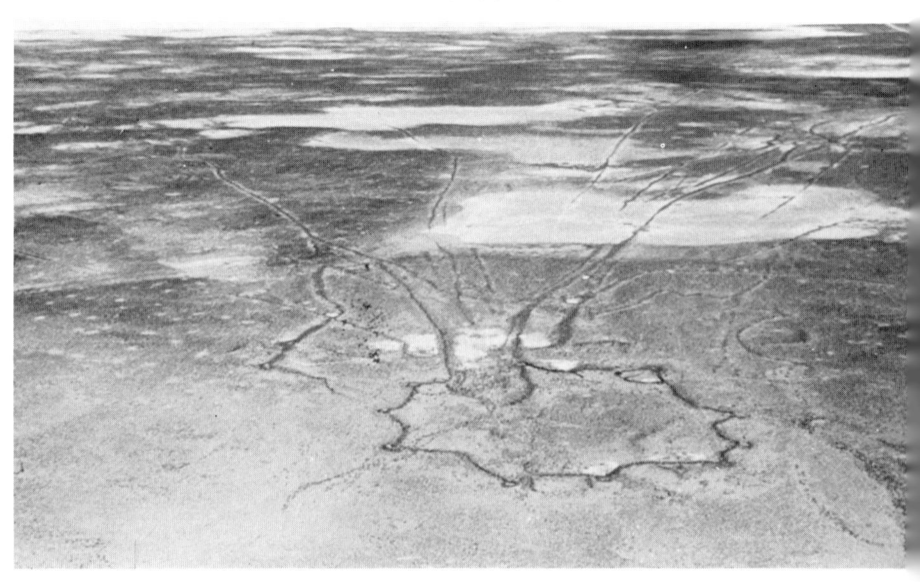

10 A 'desert kite' south-east of Jawa with clear indication of repeated alterations, not only of the enclosure but also of the radiating guiding arms (SWH)

11 The small 'kite' just south of Jawa. Its arms were later incorporated into the water systems of the ancient town

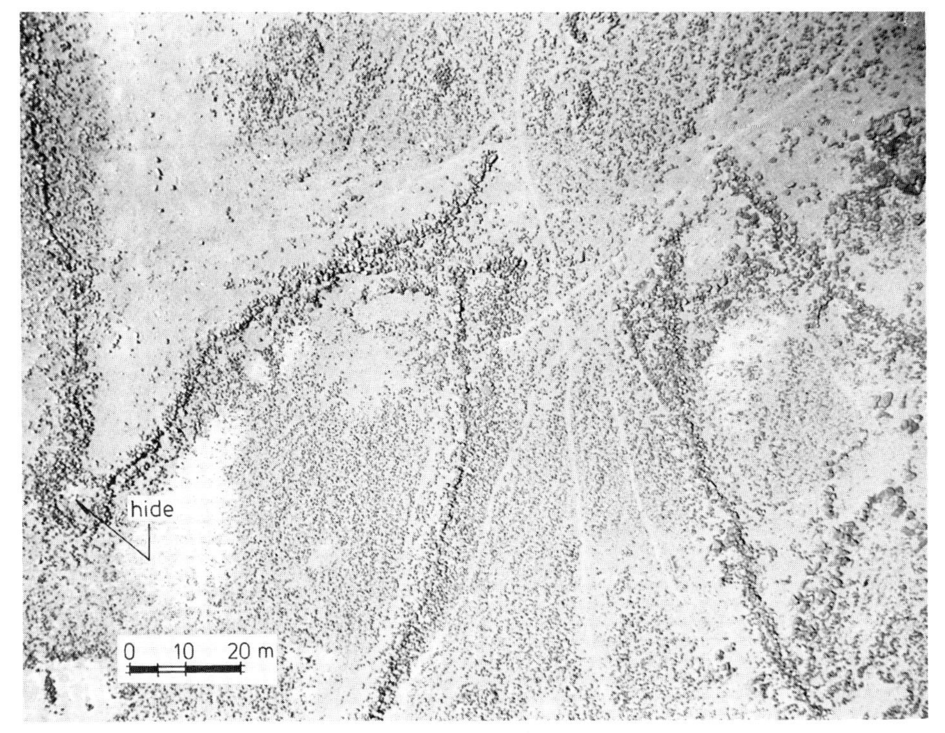

12 Detail of a typical large 'kite' showing the arms converging at the entrance to the enclosure. On the left is a hide (SWH)

13 The palette of Narmer: the lowest register depicts two captives, one representing a fortified town (or towns); the other, according to Professor Yigael Yadin, a 'kite'. Note, however, the similarity between this last symbol and the reeds in the register above: symbolic of the Nile Delta (SWH)

14 The wadi Ghadaf stone: Safaitic inscriptions and a rare drawing of a desert dwelling with animal pens (LAH)

15 Jawa from the south: virtually indistinguishable from the natural basalt formations on surrounding hills; the Jawaites' first view of their new home

16 A cave beneath the upper town: now a sheep pen, once the first home of a Jawaite family

17 Area C trench II: the first phase of Jawa; a small bin and some ashes on bedrock

18 The subsidiary valley west of Jawa in which lay two natural pools watered by run-off from the land to the west (SWH)

19 Jawa in 1975: the long straight fortifications on the western flank of the upper town, gate UT1 in the centre (SWH)

20 The massive walls of Fortress Jawa

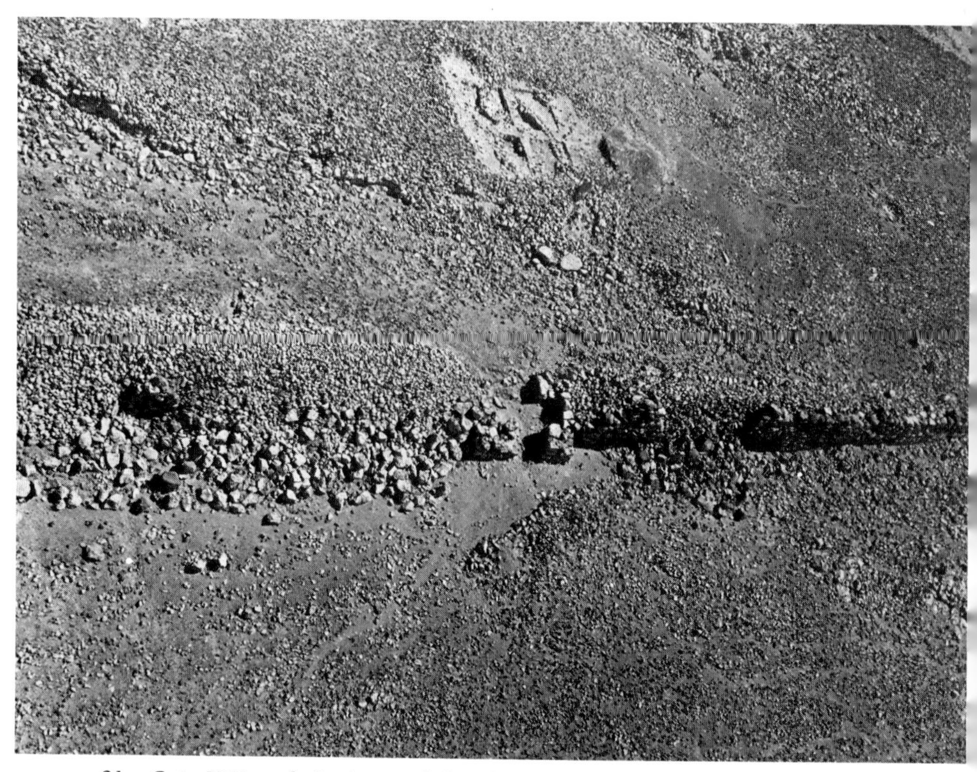

21 Gate UT1 and the internal fortifications built at the end of phase 2 (SWH)

22 Gate UT1 : eastern chamber masonry

23 Gate UT1: view from the ramp after clearance in 1973 (LAH)

24 The west wall of the upper town showing the rubble core between coursed faces

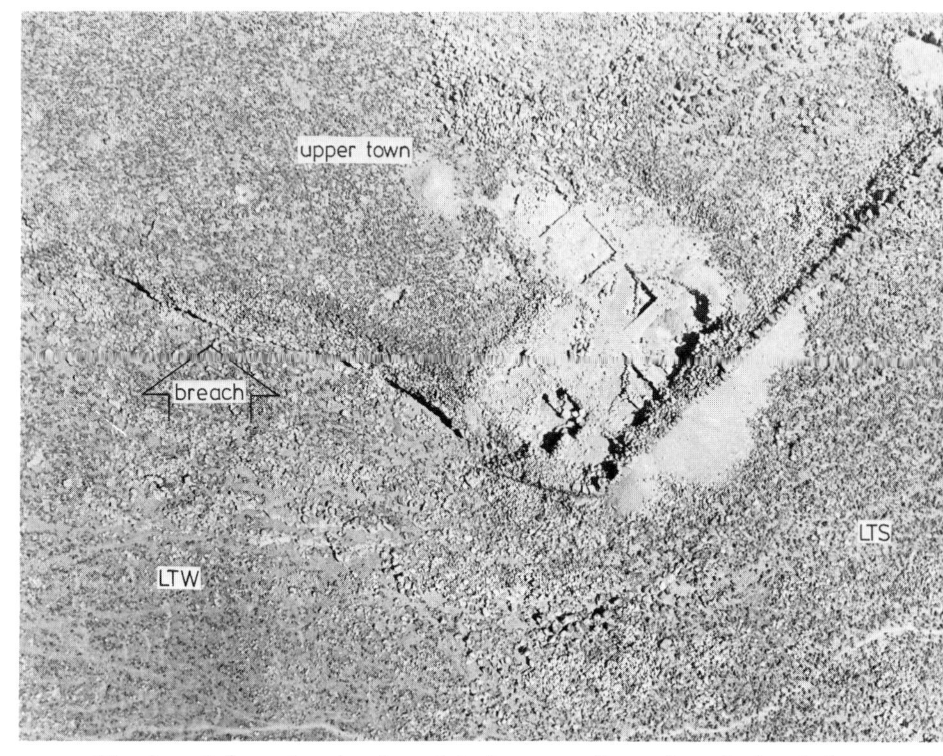

25 Area F from the air: the original west wall was breached and then rebuilt more irregularly during phase 3 (SWH)

26 Area F: detail of the 'kiln'

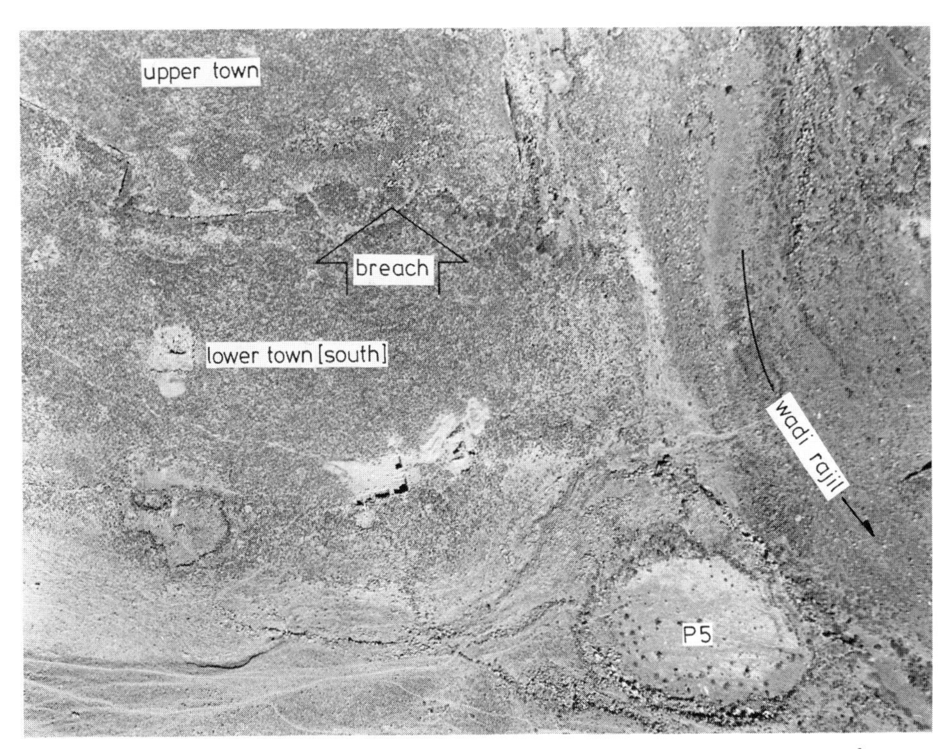

27 The south-east corner of Jawa: upper fortifications and a second major breach, area LF on the lower-town defences and pool P5 beside wadi Rajil (SWH)

28 A simple postern in the lower trace

29 Jordanian Army water tanker: the daily ration

30 Pool P3: the floor of the new bulldozed reservoir in September 1976

31 Basalt in wadi Rajil worn by pebbles and flood water (LAH)

Entrance to the lava flow cave at Mughara, now said to be the home of the hyena

33 Deir el-Kinn: a beduin pool with earth banks. Note the gully leading run-off water into the pool

34 Tell Rima: a modern concrete version of the Jawa system

35 Wadi Rajil: the deflection area DaI north-west of Jawa. Modern beduin reconstruction follows the lines of the most ancient water system (SWH)

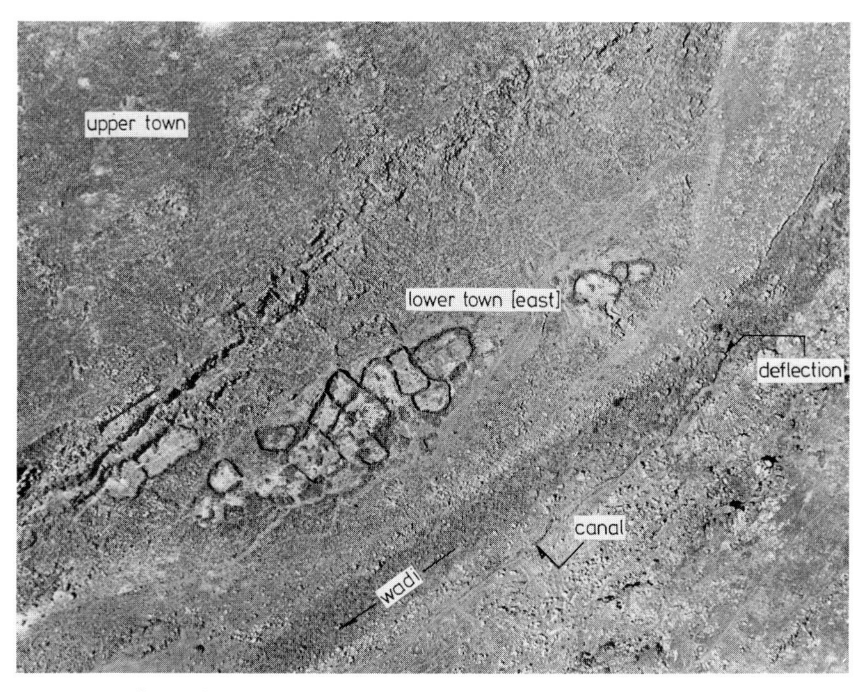

36 Wadi Rajil next to Jawa's eastern lower town: deflection area DaII leading to pools P6 and P7 (SWH)

37 Stone-lined canals near sluice gate S15 in system III. Note the recent beduin activity re-using the system over 5000 years after it was built

38 Stone-lined canal along wadi Rajil

39 The municipal water supply storage area at Jawa: pools P2, P3, P4 and the large dams (SWH)

40 The eastern end of dams D1 and D2 (SWH)

41 Dam D2 (failed)

42 System II and canals leading to pools P6 and P7. The second failed dam (DX) across wadi Rajil is in the foreground on the left (SWH)

43 Dam D2: sondage to bedrock

44 Dam DX (failed)

45 System III: pools P8, P9 and P10 and the canal from deflection area DaIII. The animal watering points were not altered during the last phase at Jawa (SWH)

46 The oasis of el-Azraq: perennial water on the road to the west . . . to the Promised Land (SWH)

wadi Rajil is the principal. But not all of the catchment precipitation reaches Jawa. A good percentage of water percolates through the fissured basalt sheets to underground storage in aquifers and further down to move as groundwater flow well out of reach of man. There are also other losses, notably storage in the form of snow on the mountains (some ultimately reaching Jawa as snow-melt or delayed run-off) and the interference of vegetation (interception) which causes water to be returned to the atmosphere. This together with soil moisture loss through plants is called evapotranspiration and is more or less irrelevant at Jawa, applying only to the upper reaches of

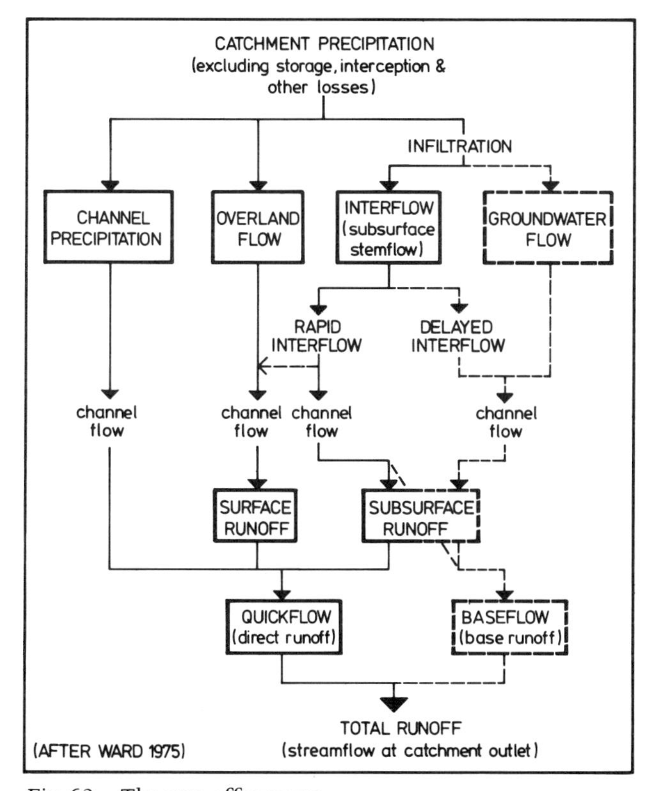

Fig. 63 The run-off process

the catchment where appreciable vegetation exists today as it did at the time of Jawa. In any case the amount lost in this way is insignificant in relation to the overall discharge of water at Jawa.

Such losses apart, only one aspect of sub-surface movement of water need concern us here and that is the result of infiltration into the shallow soils overlying the basalts. This concerns lateral movement through these upper levels until water breaks out at gullies or channels to continue downhill as channel flow. Such interflow is the only significant process that can delay movement of water appreciably and will be of specific relevance in discussing the smaller catchments near Jawa.

The soil cover, as noted, is very shallow for most parts of the region and of a type that causes desert crusting which affects infiltration. The rate of infiltration of water into soils is a function of soil particle size: directly proportional to the aggregate surface area and inversely proportional to the diameter of soil grains. The initial rate of infiltration into dry soil will be high; but in our region this decreases rapidly as a film of water forms around each soil particle and the remaining spaces are filled with additional water. Desert crusting occurs when the colloidal and very fine particles at the surface swell up and clog, causing further water to flow along the surface as overland flow or run-off. Thus within a very short time the infiltration rate either approaches a constant or, in the case of crusting, almost zero. For storms of short duration even the low constant rates are virtually negligible.

Experimental data from Jawa compare well with those derived from much more intensive tests in the Negeb Desert (Evenari and Tadmor 1971). The infiltration rates there are described in the equation:

$$f = f_c + (f_o - f_c)\, e^{-kt}$$

where 'f' is the infiltration rate in millimetres per hour at any time, 'f_c' the final rate, 'f_o' the initial rate, 'e' the base of natural logarithms, 'k' a constant and 't' the time in hours. From the comparable data comes the equation used at Jawa:

$$f = 2 + 16\, e^{-0{\cdot}65t}$$

Thus, for example, a period of rainfall lasting a quarter of an hour will produce an infiltration rate of 15·6 mm per hour and 3·9 mm of any precipitation will stay in the ground; the rest – providing the intensity of rainfall (millimetres per hour) is sufficient – will become overland flow which then moves across the surface of the ground at

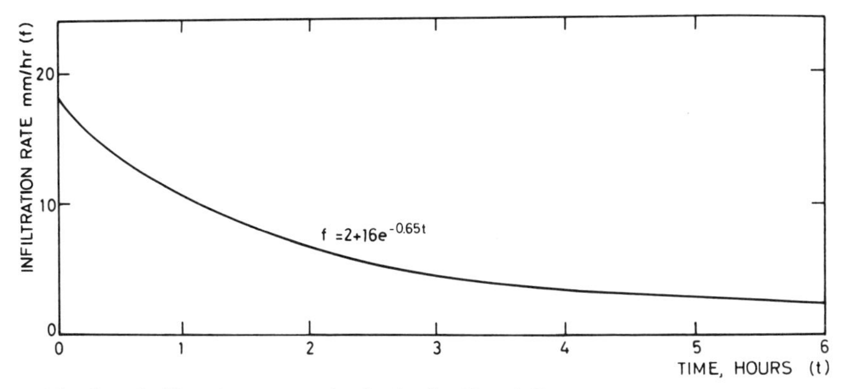

Fig. 64 Infiltration rate at Avdat in the Negeb Desert

various speeds depending on vegetation, slope and surface cover. In the Negeb Desert it was found that stones had been purposely removed and placed in neat heaps to augment the run-off rate and thus the water yield from Iron Age micro-catchments. The water continues on its way towards channels and gullies either as sheet flow covering the entire surface of the ground in extreme cases – normal to Jawa – or in a multitude of small rivulets and trickles until it reaches larger channels and finally the main arteries of the catchment wadis. Once there and of course still subject to the normal losses its rate of movement becomes subject to slope, channel dimension and configuration, and the presence or absence of erosional materials. Water now flowing down the main wadi systems (lower-order stream) eventually comes to the main artery, wadi Rajil, and in the end it is this dominant conveyor of the catchment that discharges enormous amounts of water past the site of Jawa.

Direct rainfall on the main wadi as well as on any open storage lakes that might have been built should be noted – but also understood as a very minor source of water in the run-off process in which, as has been seen, overland flow is the dominant one resulting in direct run-off (quick-flow). The amount of water reaching Jawa is vast, but the way in which it passes the site, the high order of flow rate and the timing in the wet season, will be seen as a challenge. Undoubtedly this phenomenon tested Jawaite technology to the extreme, over 5000 years ago.

Let us now look at the catchment more closely: or rather, the catchments, for there are two separate systems that once served

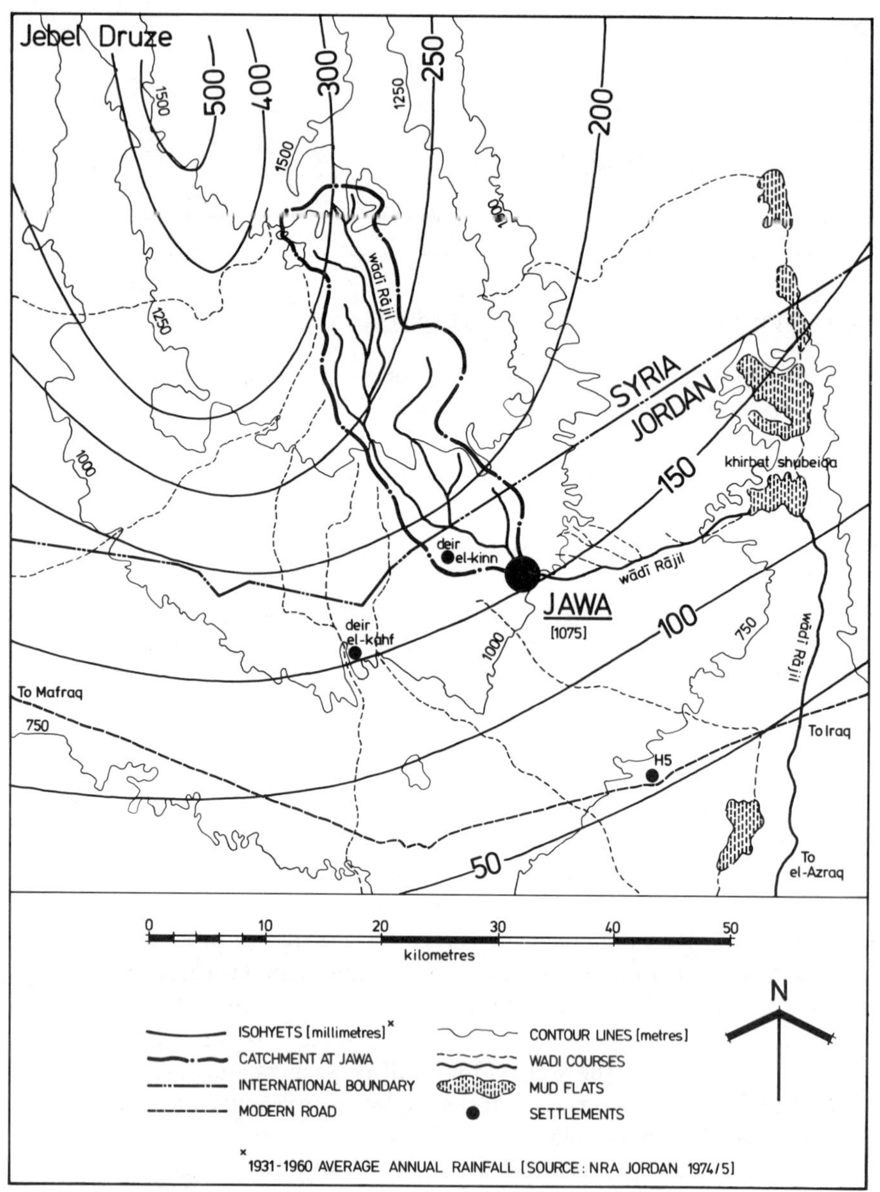

Fig. 65 Map: wadi Rajil catchment

Jawa's water needs. One, wadi Rajil, is very much larger than the other which is the aggregate of small local catchments that were eventually linked structurally at Jawa during the 'classical' stage of the town. This was done to reduce the adverse effects of fickle weather and it will be seen how precarious the balance between life and death then was.

The wadi Rajil Catchment stretches about 35 km northwards into Jebel Druze, has an average width of about 8 km and an area of *c.* 300 km². There are various catchment factors such as overall shape and channel frequency, geomorphological matters that affect run-off and ultimately discharge from a catchment that need not be considered in detail since even the lowest predicted amount of annual discharge is vast in comparison with the needs of Jawa. From rainfall data supplied by the Jordanian Natural Resources Authority

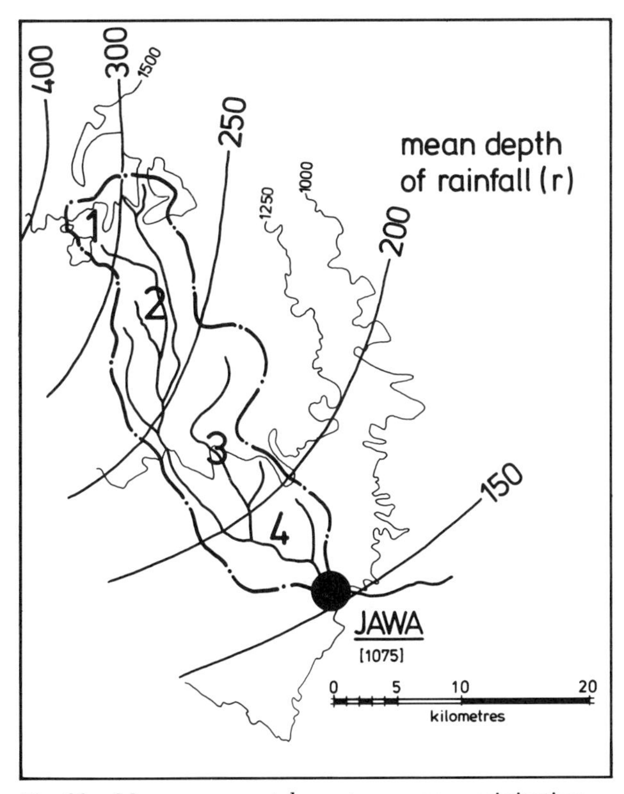

Fig. 66 Map: macro-catchment average precipitation

(NRA: Water Resources) the average precipitation onto the macro-catchment can be calculated and from that can be derived the volume of water that enters the system each year today. The simple isohyetal method has been used to calculate mean depth of rainfall between isohyets (lines of equal rainfall) according to the following equation:

$$r = B + \frac{i}{3} \frac{2a+b}{a+b} \qquad \text{(Ward 1975)}$$

where 'r' is the mean depth of rainfall between isohyets, 'b' the length of the lower value isohyet (B), 'a' the length of the higher value isohyet (A) and 'i' the isohyetal interval (A − B). The value of 'r' for each interval is then weighted for area, the whole summed and divided by the total area of the catchment to give the average annual precipitation onto the wadi Rajil Catchment. This comes to 243·7 mm, which represents over 70,000,000 m^3 of water entering the system in an average year. Of this, limiting factors understood, as well as others to be noted later, by analogy with more intensively recorded but comparable systems in Jordan and elsewhere, one may predict that only 2,000,000 m^3 of water would reach Jawa.

The micro-catchments about Jawa, one of which (C3) was the source of water that the Jawaites probably found upon arrival, are given now, although they were not completely exploited and linked in a grand scheme until the next stage. In the calculations comparable data from other areas of Jordan (source: NRA, Amman) and from the Negeb Desert (Evenari and Tadmor 1971) have been referred to more than once, the latter source being very useful with regard to micro-catchment run-off yields in relation to rainfall and the various ground conditions. With the usual qualifications entailed in comparing two separate regions on the basis of experimental data, use has been made of the nomogram published by Evenari and Tadmor showing the rainfall-run-off relations at Avdat: the effect of size, slope and surface cover of the micro-catchment on annual run-off, the scale of rainfall being 0–150 mm. Thus the catchment C3 with an area of *c.* 115 hectares, an average slope of 2·6 per cent and minimal stone and vegetation cover, might generate a run-off rate of 150 m^3 per hectare, giving a potential annual yield of 17,250 m^3 under ideal conditions, which of course had a very small chance of ever occurring. That would depend on whether a particular set of storms passed this specific point and the process was repeated several times throughout the wet season. Furthermore these storms had to oblige and discharge their water then and there. Thus the true yield is probably lower and these calculations useful only in a very theoretical way: for example, to be set against the maximum natural storage capacity at Jawa, taken as 1800 m^3, prior to the formal construction of reservoirs. Locally generated run-off, therefore, but very little else, would have to be sufficient most of the time to provide such a supply.

How fickle the weather can be in these desert regions has been described often. It is for instance a very common experience to watch an isolated storm passing back and forth, pouring rain on but a narrow strip of the desert. One can stand within metres of such a deluge and remain dry. Furthermore, so far as run-off is concerned,

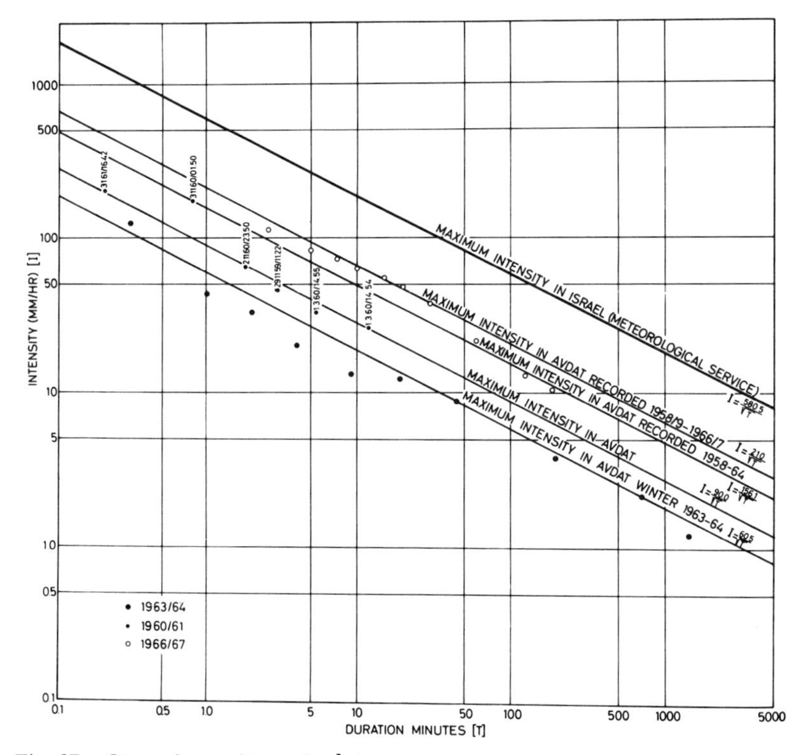

Fig. 67 Storm intensity at Avdat

even if the rain manages to be on time and score a direct hit on the relevant catchment, much depends on the intensity and duration of the storm. The Negeb data once again provides us with a parallel where the relation of rain intensity and duration has been expressed in the equation:

$$I = KT^{-\frac{1}{2}}$$

where 'I' is the rainfall intensity (mm per hour), 'T' the duration of the storm in minutes and 'K' a constant representing frequency. The lowest rate cited in the Negeb examples has been used, giving 'K' the value of 60·5. For the micro-catchment in use at Jawa during phase 1, a storm lasting from two to ten minutes would result in negligible run-off; a storm of one hour (I = 7·81 mm per hour) would perhaps yield up to 700 m^3; and a storm lasting twenty four hours (possible but very rare) would produce about 38·2 mm of rain resulting in about 2500 m^3 of run-off.

One can easily appreciate how precarious a situation was faced by the Jawaites. In order to fill the meagre natural pools even when they were revetted quite a number of lucky storms would have to break exactly at the right time and last a statistically improbable time.

18

WADI RAJIL: THE MACRO-CATCHMENT

Wadi Rajil was the major challenge at Jawa and to the Jawaites it might have seemed like an unpredictable, monstrous cornucopia that could literally drown man with plenty. We have estimated that an average discharge of about 2,000,000 m^3 of water could be expected each year and that there is no reason to doubt that a similar pattern and order existed in ancient times. Obviously fluctuations occurred and these will have to be considered in their turn, but the actual amount of water that passed Jawa each winter need not be questioned. One has only to observe the difference in order between the wadi Rajil discharge and the average amount of water (less irrigation needs) that was passed through the water systems: just over 70,000 m^3 or under 3 per cent of the total discharge in the wadi. Thus it is not the annual quantity of water that must be studied but rather the nature of its delivery at Jawa. In this will be recognized what is truly new in the Jawaite technology in comparison with earlier practices in the region.

Floods in deserts of the Near East are well-known phenomena and in most regions, as at Jawa, the only source of water. The volume generated depends on the amount of rainfall, its intensity and duration and on the character of the catchment. All of these factors, because they pertain to deserts, cause rapid run-off when compared to more gentle environments – and this run-off quickly turns into violent flooding each season. While flooding is a disaster in other regions it is the expected and often hoped-for norm in the desert. The destructive force of these norms can be appreciated from the following account concerning the Negeb Desert:

> In 1937/8 about 100 mm of rain fell in the vicinity of Auja, instead of the 73 mm customarily received during the whole year. This downpour resulted in a flood that inundated large areas and caused considerable

damage .... Not only do heavy rainfalls cause such destructive floods, but an accumulation of trash and erosional debris within the wadi also make for equally destructive torrents, often under a blue and cloudless sky. Such debris is caught up in an advancing wave and serves as a moving dam. The water stage builds up behind the obstructive debris and, sooner or later, virtually explodes under a growing hydrostatic head. These explosive releases of heavily bulked floods are especially destructive. It is no wonder, then, the Bedouins . . . said of Wadi Hafir . . . should a *seil* (flood), once come down the Wadi ... there would be an end of all prosperity in the land. (Mayerson 1960)

At Jawa things were a little different. The hope was that a *seil* would come; but the potential violence could nevertheless be a problem. This aspect of desert hydrology was horribly demonstrated some years ago near Petra in southern Jordan when a group of tourists, disregarding warnings, was overtaken by a flash flood in the narrow chasm near the ancient site. Their bodies were so mutilated by the time their parts reached wadi Arabah that only dentists' records could identify them.

To illustrate the variability of flood intensity we can look at two recorded occurrences, again in the Negeb Desert area, that demonstrate the diurnal as well as the annual distribution of flood waters. The examples both include catchments comparable to Jawa's macro-catchment and although the rainfall is much lower than on Jebel Druze the behaviour of floods is somewhat similar.

Wadi Abyed (catchment area 200 km^2) flowed six times in 1956–7 in two large and two small floods of which three produced a discharge of 50,000 m^3 each. One flood is particularly significant. It came on 9 February 1957 and it is said that 200,000 m^3 of water flowed in twenty four hours. The flood lasted seven to eight hours, stopped, and then went on about eight to ten hours more. The second example comes from wadi Asluj (catchment area 250 km^2) and there the figures are:

Year	Rain (mm)	Discharge	Individual floods	(m^3)
1943–4	79·4	431,000	16–17 April	183,000
1944–5	152·6	8,000,000	May	6,000,000
1946–7	96·9	500,000	January only	–

Mayerson, who is quoted here, considered these figures high, particularly those for 1944–5, and that seems likely in view of the exceptionally high run-off rate they imply. But even with adjustments the point is made regarding annual variation. Further, that year (1944–5) the total discharge was quite evenly divided in time (if not volume) coming in mid-November, end of December, mid-January, end of February and mid-May. Notable is the amount for May. Mayerson has summarized the implications of these floods as follows:

1 Winter rains are so capricious that one wadi may have a large flood while another, a few kilometres away, may have only a small flood or none at all.
2 As a rule, small wadis run every year.
3 The distribution of floods within a year is variable; some may come at the end of the growing season.
4 Considerable quantities of water are available for short periods of time.

Most of this applies to wadi Rajil in principle; however, the quantities involved are larger and the variability in annual discharge smaller because there is more potential precipitation on Jebel Druze. Less chance therefore exists that no water at all or merely a little flowed in any one year. But there is (and was) every chance that too much came, too fast and at the wrong time. This is simply a fact of the run-off process in the wadi Rajil Catchment with its countless tributaries, steep descent and bare rock, all causing water to accumulate rapidly in the main wadi. It is also the nature of storms, as much as the physical characteristics of the catchment, that determines the events at Jawa.

Hypothetical storm characteristics have been variously illustrated and one such 'model' (Ward 1975) has been adapted here to the wadi Rajil Catchment. The slight chance that all storms will miss the catchment has of course had to be ruled out. Thus a storm advancing on Jawa along the length of wadi Rajil (situation B) would first cause run-off in the northern regions, furthest from the site, and this water would gather *via* the many gullies and tributaries in the lower order streams (V, IV), increasing in volume at their confluence and augmented by similar flow from other lower order streams (III). By this time quite a sizable flood would be coursing southwards along upper wadi Rajil, building a hydrostatic head like a small tidal wave below the confluence of stream systems V, IV and III. Simultaneously the same process would be taking place in stream system II and this water too would be added, swelling the racing torrent almost to breaking point. Still more water would join from stream

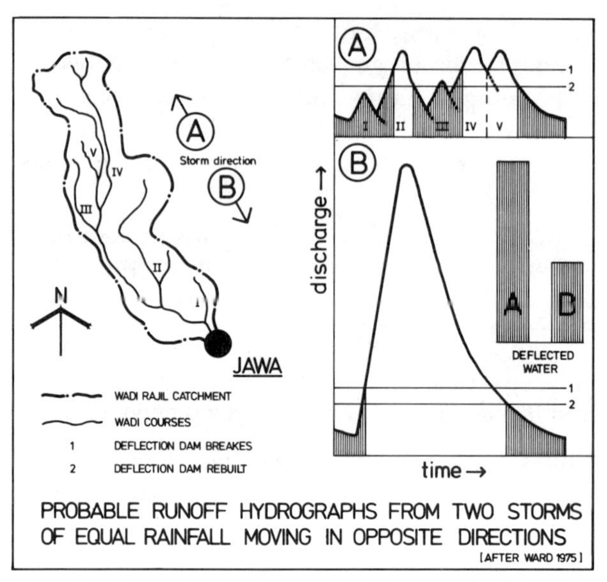

Fig. 68 Hypothetical run-off hydrograph for two storms

system I and this mass of water, choked with erosional debris, would come crashing down the wadi, swerving in the wide curves of the valley past Deir el-Kinn, spilling the banks below Jebel Haba and down the final run to Jawa carrying everything before it that was not (literally) welded down. Just short of Jawa the proof of this power is the large box canyon and the two wide water-cut pools below its 10 m-high waterfall.

In more sedate terms this phenomenon is shown in the probable run-off hydrograph of figure 68, the rapid rise of the curve reflecting the kinetic energy as well as the volume of the flood. Because of the speeds involved the peak would be reached quickly and the whole event finish abruptly; again as indicated by the steep return of the curve. This explosive energy, set against a man-made obstruction meant to stop the rush of water, can leave no doubts as to the outcome. Any dam, no matter how strongly built, would almost immediately disintegrate and a negligible amount of water could be controlled for but a tiny moment on either side of the peak flood. The time just after the peak would be useful to man only if he were quick enough to rebuild what had been destroyed.

Compare this to situation A, recalling the flood described in wadi Abyed on 9 February 1957. A storm would move up the catchment, away from Jawa, and as it passed over the various stream systems the water deposited in them would reach wadi Rajil at different times, in succession. A series of floods would therefore reach Jawa and any obstruction in the wadi. It would not all happen at once. Discharge would be intermittent and even if the yield from two stream systems overlapped the flow rate could never be as high as in situation B. The hypothetical hydrograph might appear as shown here and a dam, even if it was partly carried away, could have been rebuilt: simply because there was more time between the substantially lower peak floods.

It can be seen that the potential water harvest from floods by deflection into canals on the wadi bank is vastly different in these two examples and entirely dependent on something as chancy as the direction of a storm. It can be appreciated also that no permanent dam which did not allow most of the water to pass over it could survive in wadi Rajil. No useful proportion of the annual catchment discharge could be detained at Jawa in the main wadi: not even the 3 per cent suggested as the estimated amount required by the town. Everything depended on how much water could be deflected in a short time within a brief wet season. As will be seen, the actual timing of peak floods was a vital factor in the general water balance of Jawa and hence yet another chance-related variable making urban permanence in this marginal area a matter of such a delicate equilibrium.

In principle water was retained at Jawa by deflection – not total damming – and in practice this entailed the construction of dams to direct a small part of the water from the big valley into canals leading to storage areas where it had to last through the long dry season. This required a knowledge of weather patterns as well as hydrodynamics, surveying, earth mechanics and masonry. It required above all an understanding of science in terms of observation, recording, evaluation and prediction. Much, if not all of this work had to be done in a very short time between the arrival of the Jawaites at the site and the next rains. There was no time for developing since there was really no time for experimentation. Science is not simply invented; and this is the implication at Jawa. The water systems are a summary and testament of the Jawaites' achievement. The schematic representation of their concept of water retention and conservation now to be described only serves to underline this by its complicated and almost modern orderliness.

19

THE WATER SYSTEMS

Water from wadi Rajil was deflected at three points and linked to three storage areas by a network of gravity canals. Of these three systems the biggest undertaking, as one would expect, was the one serving the 'municipal' water supply.

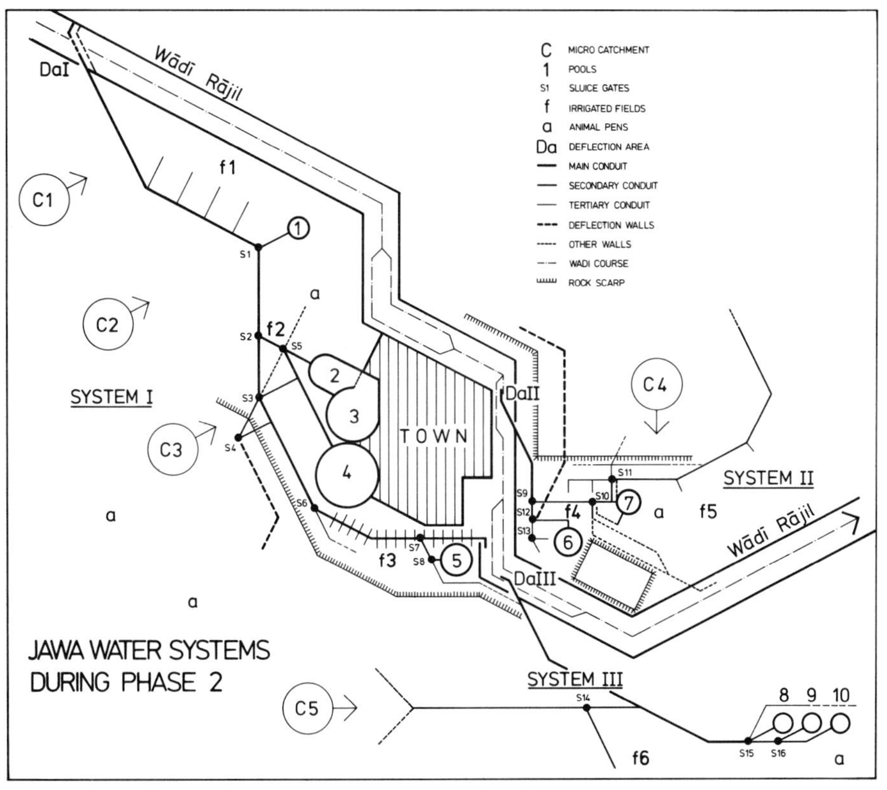

Fig. 69 Schema of the Jawa water systems

System I consists of a canal leading water from the deflection area (DaI) past a series of irrigable fields to a sluice gate where some flows into an underground cistern (P1). The main canal continues to another sluice gate where it divides to serve the town's drinking water containers, P2, P3 and P4, next to the western lower town, passing through a second area of fields. The second branch by-passes the storage area, acting as a spillway, and passing through or along a third system of fields reaches an animal watering point, P5. Surplus water is returned to wadi Rajil.

All along the gravity canal above sluice gate S2 additional run-off is captured in linked micro-catchments, C1, C2 and C3, the southernmost of which is provided with a deflection wall along the edge of the cliff overlooking Jawa from the west. This wall extends the catchment area considerably and leads water either along the by-pass canal to pool P5, or to reservoirs P2, P3 and P4. Micro-catchment C3 was the first exploited at Jawa and it is quite possible that at least catchment C2 was also harnessed at that time since the gravity canal lies in a natural gully draining into the main municipal storage area.

System II consists of the deflection area DaII opposite the eastern quarter of the lower town and raises water along a canal sited at the bottom of the steep eastern shore of wadi Rajil. The canal leads to a sluice gate where water could be distributed to pools P6 and P7. As in system I, a micro-catchment was incorporated (C4) and likewise increased in area by a long canal enclosing the lower slopes east and north and a stone deflection wall at the western edge along the cliff of wadi Rajil. Run-off from the micro-catchment was led along the curving canal past a number of fields to pool P7. The western part of the micro-catchment drained towards this pool as well, but it appears that provisions were also made for the irrigation of fields. The two storage areas were probably animal watering points, just like pools P1 and P5 in system I.

System III also begins at a deflection area (DaIII), just south of pool P5, and raises water along a gravity canal to several sluice gates and three storage areas, pools P8, P9 and P10. To underline the homogeneity of the entire scheme, system III also has a micro-catchment (C5) and several fields associated with it. There may even have been more fields below the major canal in this system, but it is hard to be certain about this because of recent disturbances.

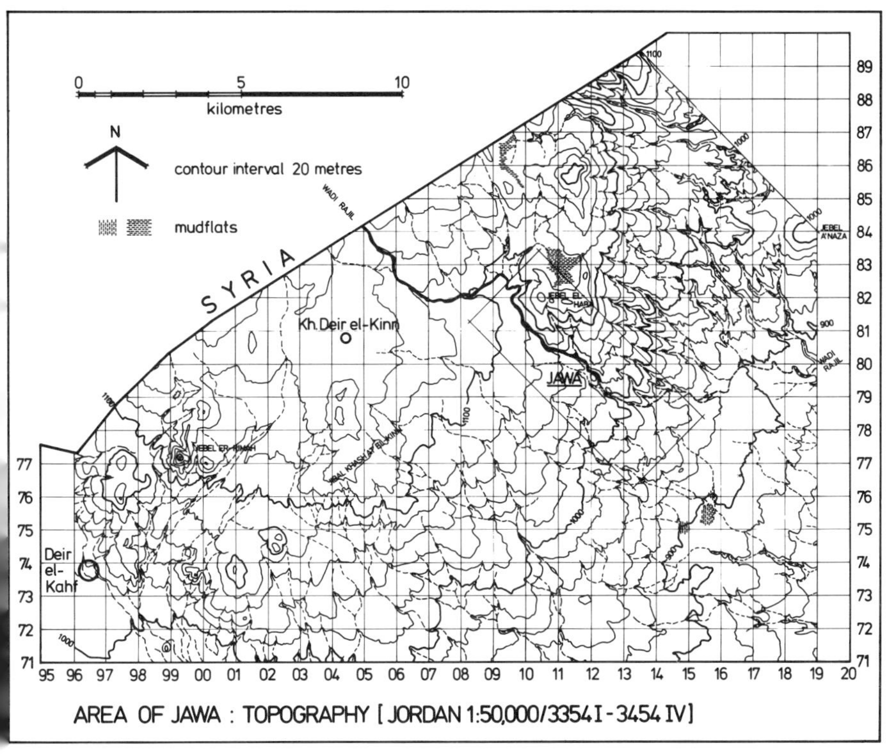

Fig. 70 Map: Jawa area topographical

As an integral part of these three systems, animal pens (a) are built over the remains of desert 'kites' to the west of Jawa and in smaller concentrations near the animal watering points of systems II and III.

This then is the general organization of the most comprehensive urban water system yet discovered anywhere for such an early date. Others undoubtedly existed elsewhere in the Near East and probably at much earlier stages in the development of the city, but they await discovery. However, unless they are sited in deserts and have not been covered with the normal millennial urban debris that characterizes tells in the Near East there is very little chance that anything as complete as Jawa will ever be uncovered. Thus Jawa is all we have and we must now consider whether all of it is really as it appears in our schematic diagram: whether all the systems and their parts are indeed one scheme and the achievement of the 'classical' stage at Jawa. In the detailed descriptions below it will be seen that there is definite chronological and structural proof of two phases related to the construction of system I. There we found an earlier dam (D1) beneath the later one (D2) still visible today. Both date from the later fourth millennium and the urban stage of Jawa. The later dam is structurally the same as yet another dam build across wadi Rajil between pools P5 and P6, just upstream from deflection area DaIII. Both dams and their planned reservoirs failed miserably and are the last water-retention structures that can be associated with a permanently-settled Jawa. These dams might therefore be linked to the final stage of occupation, phase 3. Dam D1 as well as pool P4 and its water-supply system up to deflection area DaI must belong to phase 2 and the period of optimum and successful exploitation of water resources.

That leaves pools P2 and P3, the irrigated fields and the underground cistern P1. Their date is not seriously questioned because we have found the requisite early flint and pottery material associated with them. Only their place in the sequence of events is less clear, although it must be before the final stages of the town. On the one hand pools P2 and P3 are designed and built in much the same way as pool P4, on the other hand the hypothesis has just been presented of development during phase 2 concerning the growth of Jawa's suburbs and an increased demand for water. The problem of access to the water supply has also been noted. Thus it

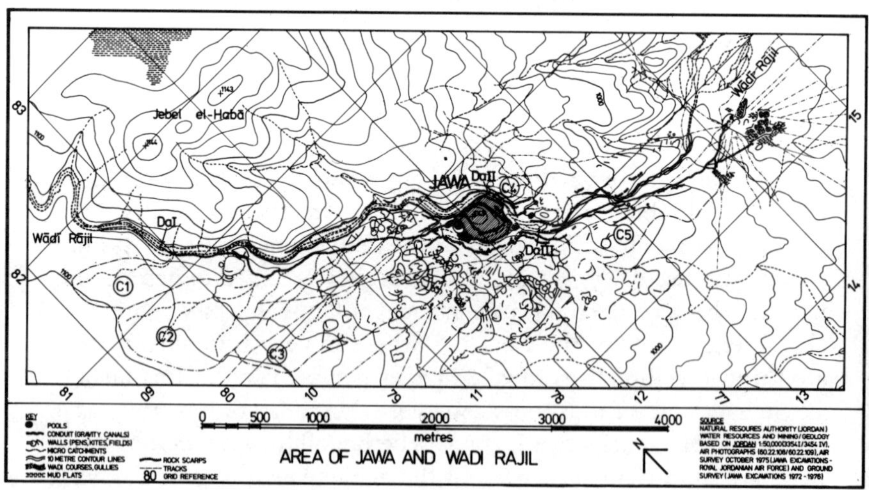

Fig. 71 Map: Jawa area detail (water systems)

can be argued that the pools were built consecutively, as they were required: an idea not incompatible with the construction timetable proposed for the period between phases 1 and 2. The general design and development of the water systems would logically be the product of the second phase and in terms of design similarity the other two systems must fit in the same way. There too we found acceptable dating proof linking them to urban Jawa. Sequential stages within phase 2 must be determined structurally and stratigraphically (where this is possible) and after that logically according to design and change in the town itself. The probable subdivision of the first two phases of the water systems at Jawa may have been as follows:

Phase 1		Pool P4, canal linking micro-catchments and a first version of DaI(i): *c.* 5000 tons basalt
Phase 2	a	Pools P3 and P5, DaI(ii): *c.* 10,000 tons basalt added to DaI(i) – Compromise 1
	b	DaI(iii) finished, pools P2 and P1
	c	Systems II and III – Compromise 2

20

DEFLECTION DAMS AND GRAVITY CANALS

Structurally the entire water scheme at Jawa is a matter of earth and stone. The low terminal infiltration rate into local soils makes them ideal fill for gravity dams. Stone provides the form and counters wave action and scouring in the reservoirs. In discussing the construction of the systems let us first look at the deflection areas.

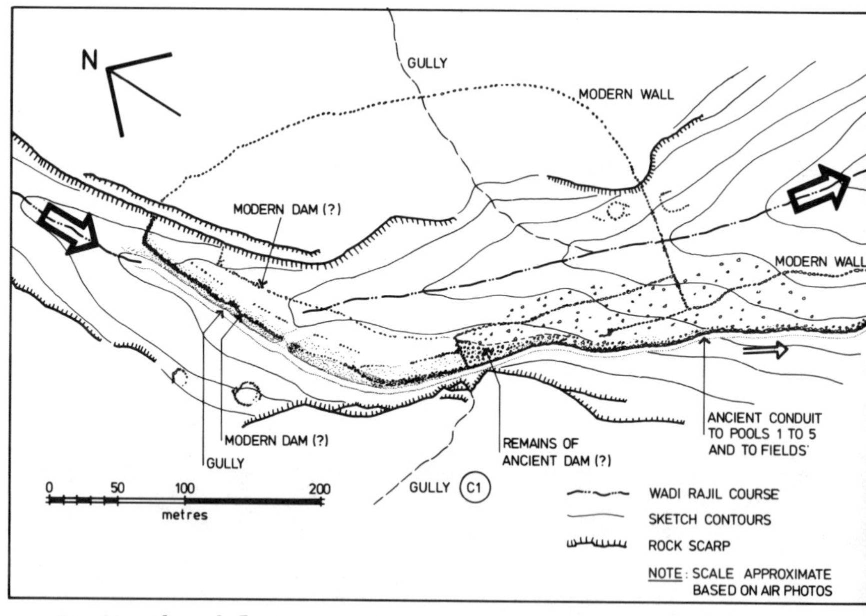

Fig. 72 Plan: deflection area DaI

The choice of site was limited in area DaII which had to be where it was. The others, however, show that the Jawaite engineers understood the power of moving water and the hydrodynamic

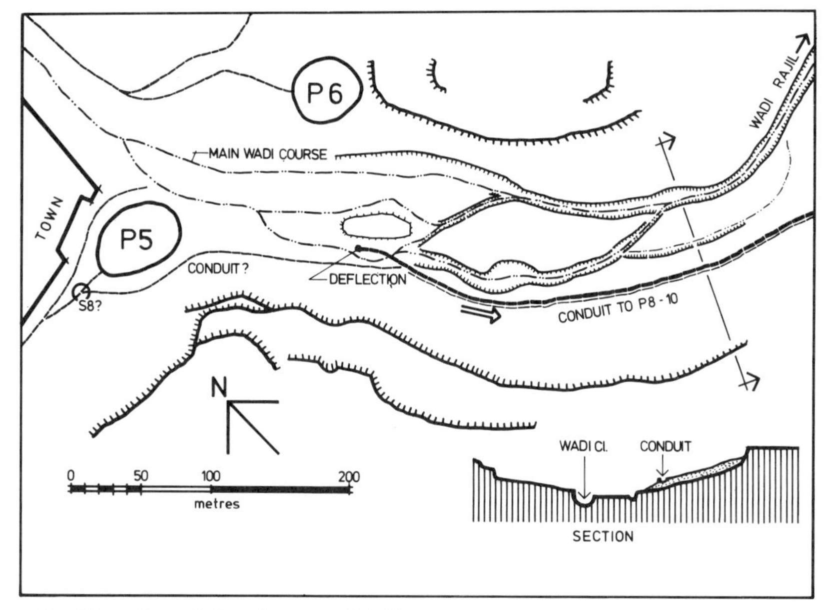

Fig. 73 Plan: deflection area DaIII

principles involved in deflection. In area DaIII it is clear that whatever obstruction was built in ancient times, long since washed away by the wadi, it was placed across a subsidiary stream bed, thus reducing the chance of destruction as much as possible. Area DaI is obviously the most important deflection of them all.

With the aid of ground and aerial surveys it is possible to reconstruct some of area DaI. This is not easy since nearly all the ancient walls have long since vanished from the wadi bed and more recent ones have been built re-using the old system. The modern work, however, does at least show how deflection was managed. The choice of site is obvious, as are some of the limitations. The latter concern absolute heights and the box-canyon further downstream. Deflection had to occur above the 'waterfall', but in terms of slope not too far upstream. There was still some choice and we know that the Jawaites tackled the matter successfully: they apparently understood to some extent the multi-dimensional phenomena of channel-flow, meandering and the complex inter-relationship between hydrological and geomorphological variables. They seem to have chosen precisely the correct spot in the wadi, at a point where it changes direction to the left (looking downstream) and where water would riffle, lose energy and tend to mass towards the right bank where (indeed) clearly modern structures merge with remnants of the fourth millennium gravity canal.

Since only the line of the ancient dam remains today its original structure must be an inspired and hopefully realistic guess. There is of course the obvious contention that something must have been built here in wadi Rajil: otherwise, no Jawa. We have the modern structures which still diverted water until recently to guide us. There are also some ancient clues.

At the mouth of the gravity canal one can see that the stone revetment on the wadi side was widened. This is probably part of the original Jawaite deflection dam. The appearance of its continuation across the wadi bed is another matter. But again, it was either a partial obstruction made of basalt boulders, allowing nearly all of the floods to pass, or it was a proper dam across the whole of the stream bed along the lines indicated by the more recent walls. Only the latter requires further comment since the former is better preserved in system II.

An obstruction across wadi Rajil could never have contained all the flood water. There is neither the construction that could withstand the power of the wadi in winter nor the topography that would create a large and useful lake. Rather, such a dam would have to act as a weir or spillway, a kind of threshold over which most of the floods would pass on downstream, and only a tiny amount of water (i.e. 3 per cent) would be deflected towards the canal on the right bank. Spillways in the Near East have been described elsewhere and from the Negeb Desert come the following dimensions for albeit much later structures which are nevertheless relevant since they are made of similar materials:

1 Spillways with crests of 30–60 m withstood floods in the order of 10–30 m^2 per second (cumecs)
2 Spillways with crests of 3–5 m: floods of 1–5 cumecs.
3 Spillways with crests of 1 m: floods below 1 cumec.

The fourth millennium deflection dam in wadi Rajil would have had a crest measuring 150–200 m, set at an oblique angle to the stream bed, and perhaps a width of about 40 m. This is by analogy with better-preserved dams at Jawa. It would be revetted with roughly-coursed boulders upstream and downstream and be infilled with stone or gravel since water had to pass over it. Such a structure might withstand floods in the order of about 80–110 cumecs. However, since floods in excess of this can occur now and then and since, as noted, the wadi flows nearly 2 m deep next to Jawa (stones in hawthorn bushes), it is quite possible that part of the dam

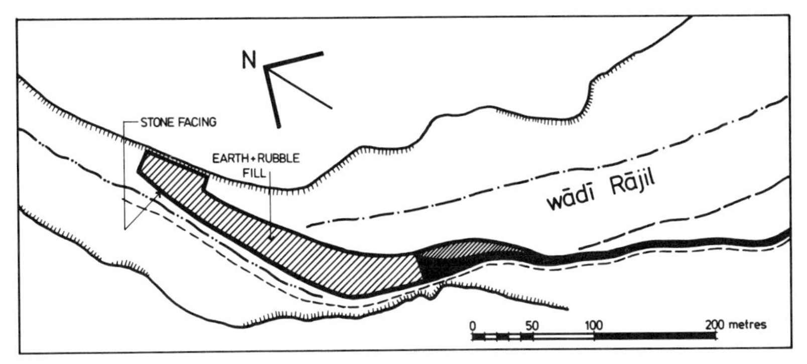

Fig. 74 Plan: DaI(iii) reconstruction

would be washed out periodically. This implies that the dam may have been designed to 'fail'. It would be in accordance with the hypothetical storm hydrograph (situation A) discussed in chapter 18 and might represent the most plausible reconstruction. The possibility of serious failure must always have been looming over the Jawaites, but as noted the systems did function for some time and the achievement at Jawa has to stand as a new – and the earliest – milestone yet discovered in man's development of hydro-technology.

In the history of technology a gravity dam of similar design is mentioned by Herodotus (Book II: 99). It was to have been built in Egypt *c.* 2900 BC:

> The priests said that Men was the first king of Egypt and that it was he who raised the dyke which protects Memphis from inundations of the Nile . . . by banking up the river at the bend [like DaI at Jawa] which it forms about a hundred furlongs south of Memphis, [he] laid the channel dry, while he dug a new course for the stream half-way between the two lines of hills.

The king also seems to have diverted water from the Nile to what might have been a municipal reservoir since after digging the new stream 'he further excavated a lake outside the town, to the north and the west, communicating with the river, which was itself the eastern boundary.

This Egyptian construction has been discarded as fictitious by historians of dams (Smith 1971) and this may well be so, especially when one considers the monumental task involved in damming a river the size of the Nile. Yet it is interesting that the principles involved in this story are very similar to those understood at Jawa several centuries earlier. Egyptian technology will be referred to again in considering other dams at Jawa, at this point it can be said that it is very probable that the hydro-technology of Jawa could easily have been known in Egypt about the turn of the fourth millennium.

It has been argued, however, that Jawa's science did not come from the south but from the north-east, from Syria/Mesopotamia. The possibility of some local invention within Arabia has also been noted. Is there any evidence, apart from the logical assumption, that necessity demanded a similar and earlier development of hydro-technology in the areas of the Rivers Euphrates and Tigris?

In Arabia the earliest known deflection dam re-routing run-off

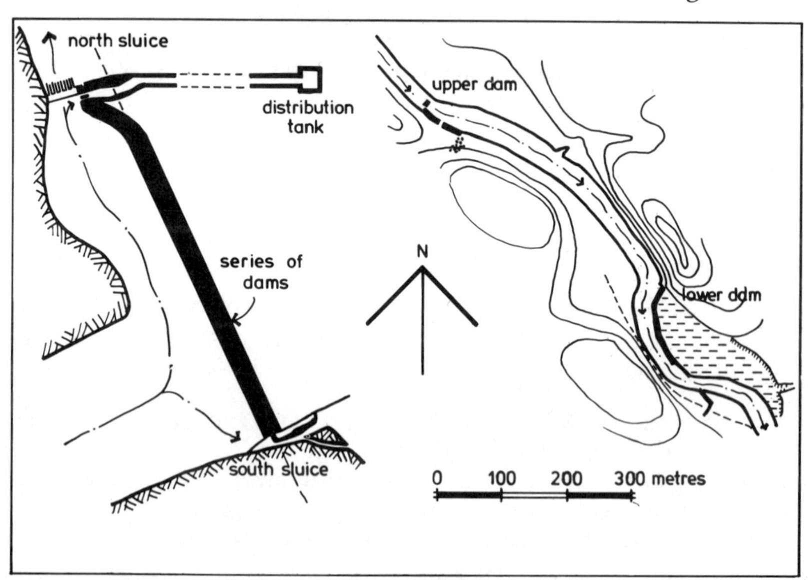

Fig. 75 The Marib dam (left) and the dams of Sennacherib on the Khosr (right)

was built many years after king Menes; even later than the next known dam in Egypt (*c.* 2700 BC). The great Marib dam in southern Arabia (750 BC–AD 575) was often rebuilt and improved (Bowen and Albright 1958). It began as a simple earth bank that was not meant to create a reservoir but rather to raise water for a system of irrigation canals serving the fields of a prosperous city, Marib of the Sabeans (Saba: the Sheba of the Bible). Thus in function and relation to run-off and urbanism the Marib dam is a direct parallel for the more ancient structures at Jawa. However, the time gap between the two dams is so vast that the evidence has little use in tracing the origins of the earlier technology.

In Mesopotamia as in Egypt irrigation and hence deflection of stream flow into gravity canals must be older than written history: indeed probably as old as the practice of agriculture. Traces of canals have been found dating from well before Jawa (Oates and Oates 1976) and it must be assumed that these were related to deflection dams of some sort. It is also understood that hydro-technology on a large and organized scale was intimately linked with urbanism, there in Mesopotamia as well as in the rest of the Near and Middle East. However, the earliest record of a dam comes from a tablet of the time of Ur-Nammu, a ruler of the Third Dynasty of Ur (*c.* 2140–30 BC) which refers to a reed dam. Another dam is attributed to Marduk, a Babylonian god of about 200 years later. South of Samarra a dam dating from the second millennium, called Nimrod's Dam, was built to divert the River Tigris into a new course. The clearest structural and design parallel to Jawa's deflection system at DaI comes from Nineveh where King Sennacherib built two dams about 694 BC to supply his city with water from the nearby Khosr river. The two dams were made of roughly-shaped limestone blocks with a vertical waterface and a stepped airface. In plan and relation to the stream bed they are very similar to the reconstruction of Jawa's main deflection dam across wadi Rajil. Both of Sennacherib's dams were designed to discharge water over their crests. The contention then is that the hydro- and urban technology evident in the achievement at Jawa was very likely known and developed long before the fourth millennium in northern Mesopotamia and north-eastern Syria. Even if there is not yet the physical proof of this one can plausibly extrapolate backwards from the dams described. This conclusion underlines the origin hypothesis for the Jawaites.

The structure of the gravity canals at Jawa is very simple and their date is given by flint and pottery scatter inside them. The construction method and design of the canals is very much within the capabilities of the pre-Jawaites of the Black Desert; but one must realize that very little more is needed to transport water overland. Even modern prefabricated concrete conduits are not so very different. They are made of better, more durable materials and set out with sophisticated surveying techniques, but no more. Jawa's canals, depending on the terrain, are revetted on the downhill side with a double stone wall, infilled with the local soils. Uphill they consist of either the natural rock or merely the earth bank of a small gully. Their path is of course related to elevation, slope and other topographical features and it seems probable that they were laid out by experiment rather than much more intricate survey – rather like the platforms that were to support the Great Pyramids some years later. They were levelled by first cutting channels into the bedrock about the periphery and using free water to indicate constant depth and hence a horizontal plane.

Since the Jawa systems are so ancient much of the original fabric has gone. This is especially true in the flatter areas. And since the systems were recently re-used additions have altered the ancient plan somewhat. Nevertheless the track of the canals is quite clear and provides us with the survey data for further calculations.

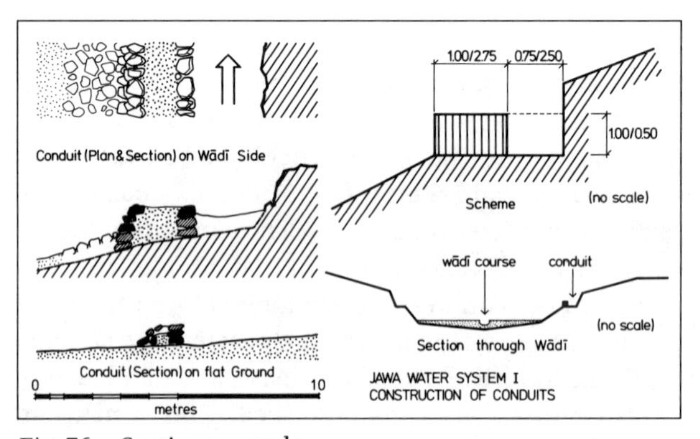

Fig. 76 Sections: canals

Table 3 System I: gravity canal

Survey point	d	Fall	Slope %	Comments
17				wadi Rajil (144·66)
16–15	135	0·710	0·53	
15–14	135	0·240	0·18	
14–13	200	0·830	0·42	
13–12	80	0·825	1·10 ?	stone-banked canal
12–11	160	0·685	0·43	
11–10	100	0·630	0·63	
10–09	160	3·430	2·14 ?	
09–08	150	1·420	0·95	
08–07	150	1·385	0·92	
07–06	230	3·120	1·36	S1 canal to P1 (131·39)
06–05	180	4·000	2.22	
05–04	240	4·290	1·79	wadi course to S2
04–03	232	5·290	2·28	and P2, P3, P4, P5
03–02	193	7·110	3·68	(103·07)
02–01	182	7·630	4·19	

In these figures the high gradient values in the stone-banked canal may be rejected as caused by recent disturbances. The relevant average gradient calculated from the remaining data shows a gradual increase from 0·4 to 0·9 per cent between the deflection dam at DaI and survey point 06 at sluice gate S1. From that point onwards the canal follows a natural gully to the reservoirs west of the town. The maximum feasible discharge rate along this system may be calculated as between 0·5 and 1·0 cumecs.

Sluice gates within the systems are as basic in design as the canals. They represent age-old irrigation practice as illustrated on the predynastic Egyptian mace-head illustrated at the beginning of this section. The direction of flow is altered simply by moving a few stones, sealing off one canal while opening the other and consolidating the 'mini' dam with a little earth. The variations in use at Jawa are shown here in schematized form and despite their simplicity their distribution within the systems implies once more the remarkable high level of organization that would be required to service and maintain them. One is reminded again of the urban

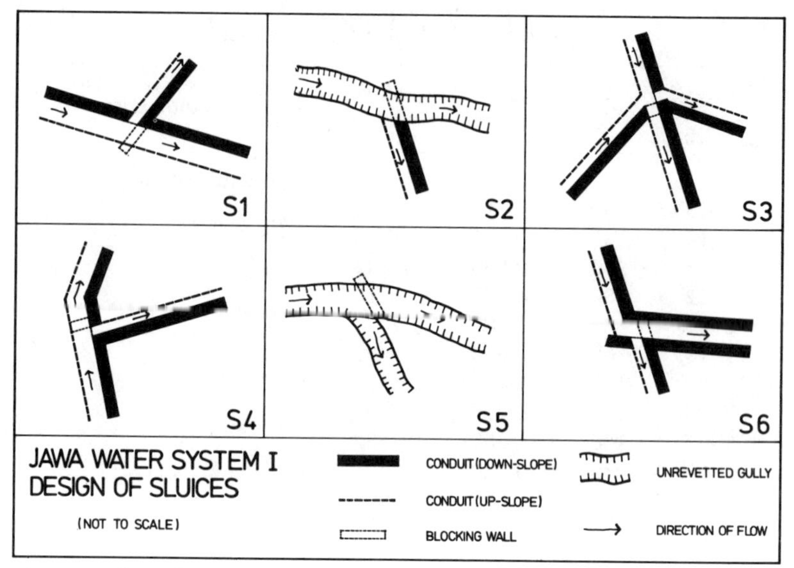

Fig. 77 Plan: sluice gates

connection of such techniques when they attain this scale – the entire water system stretches over 5 km – and with that thought perhaps the admonitions of Hammurabi are apposite. That ruler of the First Dynasty of Babylon (*c.* 1800 BC) states in his famous law code (Section 53):

If anyone be too lazy to keep his dam in proper condition, and does not keep it so; if then the dam breaks and all the fields are flooded, then shall he in whose dam the break occurred be sold for money and the money shall replace the corn which he has caused to be ruined.

21

RESERVOIRS AND DAMS

The earliest known dam before Jawa was discovered survives south of Cairo near Helwan and is called the Sadd el-Kafara or Dam of the Pagans. It was discovered in 1885 and dates from about 2700 BC. The dam was made to create a reservoir in the wadi el-Garawi by blocking the stream bed with masonry obstructions and it is estimated that about 40,000 tons of stone and 60,000 tons of gravel were required to build this monument to early hydro-technology which failed very soon after it was completed. Today all large embankment dams such as the Aswan High Dam of Egypt which created Lake Nasser are provided with a core of lower permeability: either a core-wall or a cut-off-trench running the length of the dam and filled with puddled clay or concrete. It is also common practice to consolidate the lower airface of dams by installing a drainage blanket, usually of concrete, in order to secure the structure against erosion due to seepage. The downstream sections of embankment dams are particularly vulnerable and if such precautionary steps are not taken the dams would soon melt away. It will be seen that even some of these modern features were probably used at Jawa during the later fourth millennium.

The great dam opposite the ruined town was our shelter during the first season of excavations in 1973 and we used to spend the later afternoons there under a frayed tarpaulin waiting for the sun to set. Our heads would rest against the huge basalt boulders that made up the waterface of this structure and at that time we had no idea that the yellow silt beneath us was nearly 2 m deep and that our camp would be the site of next year's sondage that was to reveal a very different story. Moreover we thought then and up to 1975 that this massive straight face was the major water-retention dam at Jawa and that certain internal walls which had come to light in the east

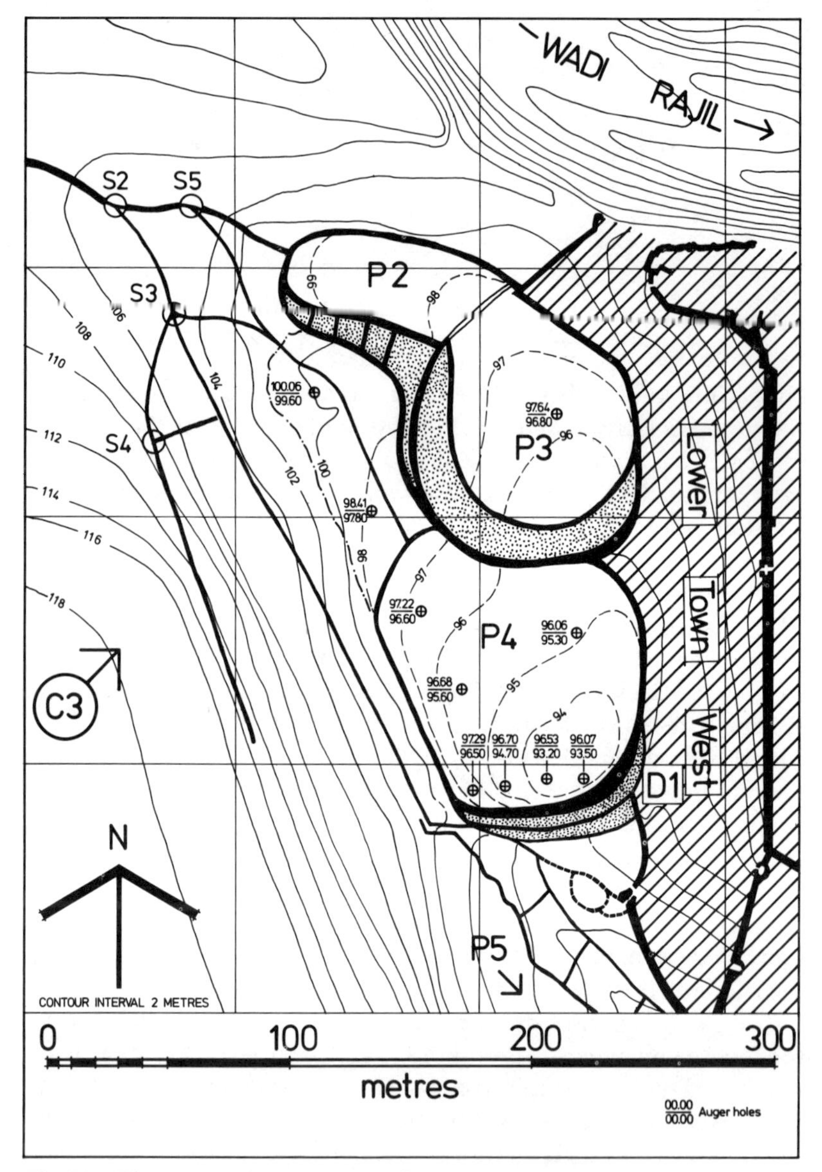

Fig. 78 Plan: water storage system I

represented some sort of spillway, a sluice gate or even a barbican associated with the fortifications of the lower town.

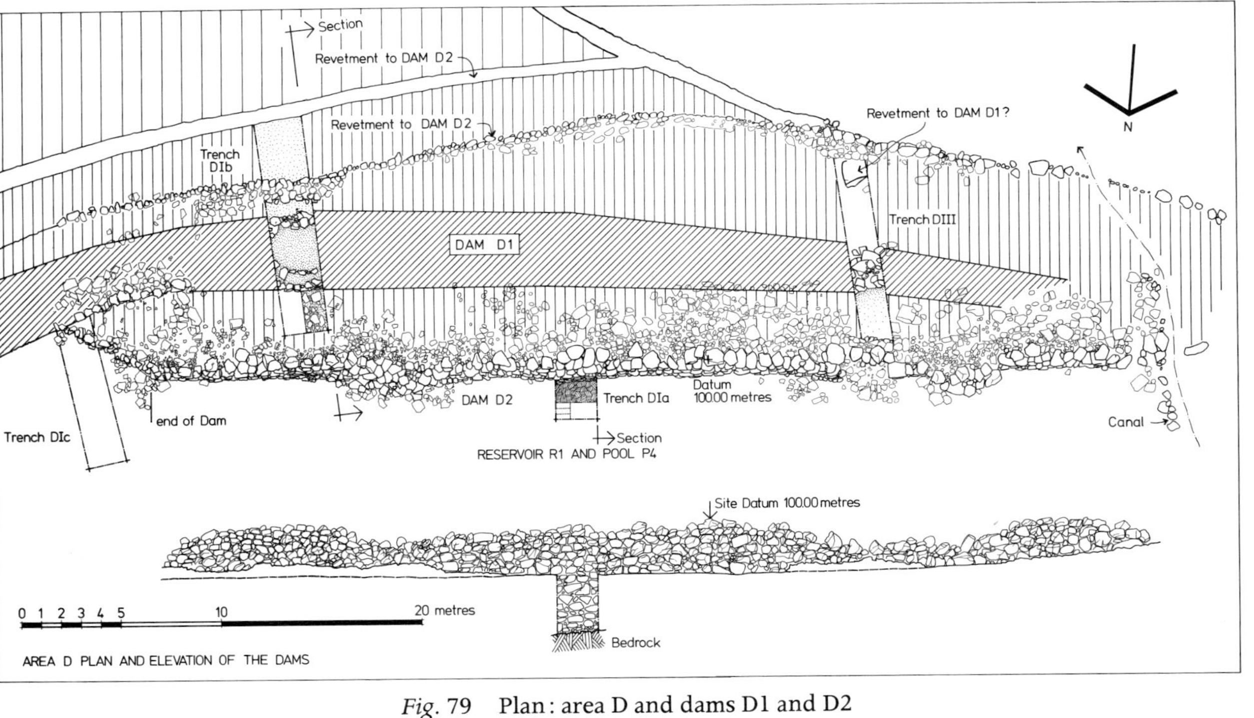

Fig. 79 Plan: area D and dams D1 and D2

In 1975 we cut trenches into the earth bank behind the face of the dam and discovered an internal structure which at first we took for a casement or internal reinforcement, perchance a true core-wall. Soon, however, it became clear that this structure was in fact an earlier curved version of the visible dam and the eastern walls first noted two years earlier merely its eastern extension. We had already associated the straight dam with the time of urban Jawa, during the previous season of excavations; now this new discovery of yet another dam – one of different design as well – brought with it the dawning of the idea that the monumental symbol of Jawa's hydro-engineering genius had never been completed and that it actually represented a sad failure. The earlier dam (D1) is the first water-retention structure here at the lower end of pool P4 and changed considerably the picture we had been building about Jawa's municipal water supply.

Dam D1 consists of a stone waterface and stone revetment downstream with an earth fill between (a primitive core-wall or cut-off-trench) made up of rock-hard clay and silt and packed with layers of ashes. An interesting and quite new feature, compared with the later dam, is the roughly stone-paved surface at the base of the waterface. This must have protected the dam from wave action at the same time as retarding seepage. In modern terms this structural addition is called an apron.

From the remaining height, here and in other trenches, it can be

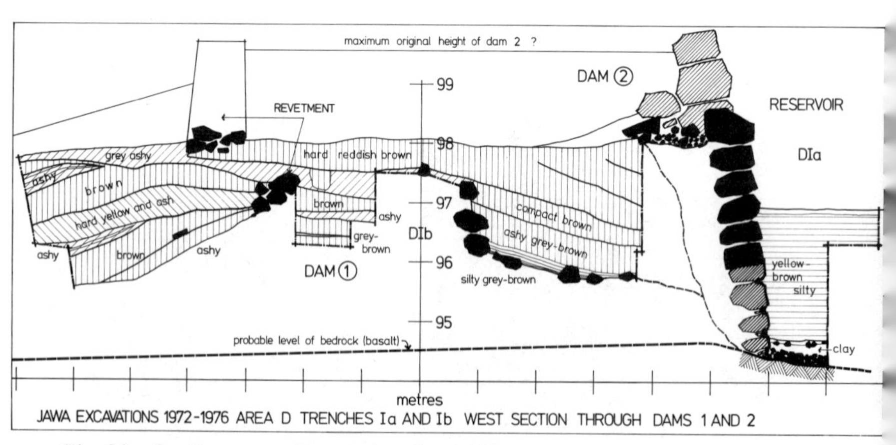

Fig. 80 Section: area D trenches Ia and Ib

concluded that the upper limit to which the dam can be reconstructed must lie about the 99 m contour line, giving a maximum height to the structure of between 4 m and 5 m at the deepest point of the adjacent reservoir and a total crest length of at least 80 m. It was probably even longer than this since the dam must have returned northwards along the western lower fortifications. Further calculations are given below. At this point it is possible to visualize the dam D1 and hence the pool or reservoir P4 which would have had an apron all around and probably an outer stone revetment (drainage blanket) swinging in a long curve from bedrock in the west to the western town wall of the lower town just north of gate LT2. In the west the by-pass canal or spillway from sluice gate S3 would lie some metres above the dam. The dam itself may also have served as a causeway to the town: a normal secondary use of such structures even today.

The material chosen for the construction and the manner of its assembly at first puzzled us. The abundant ashes in the

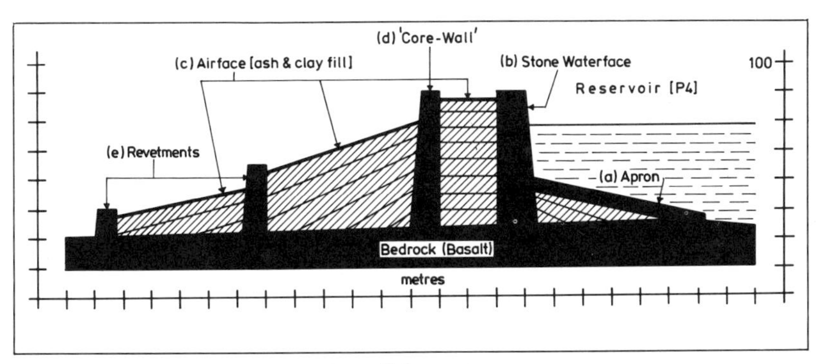

Fig. 81 Section: reconstruction of dam D1

infrastructure of the dam looked like the results of destruction rather than deliberate construction. Then it became clear that these ashes, when they are solidly held in place and compressed, are a most effective method of water-proofing; even better than clay or compressed silt which make up the rest of the core. The infiltration rate into this fill is very low and it can be assumed that losses due to seepage into the dam and subsequent evaporation through the airface were relatively slight. Such losses have been estimated to be

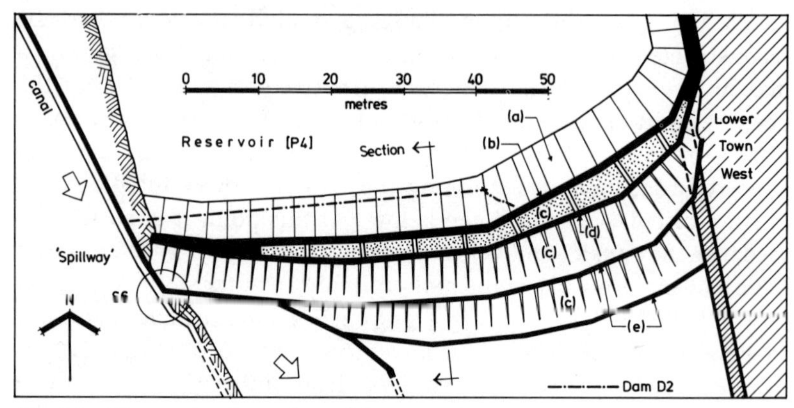

Fig. 82 Plan: reconstruction of dam D1

minimal in relation to others such as free-water evaporation. Certain correction factors will in any case be introduced in the ultimate calculations of the water balance at Jawa. The general structure of the dam, apart from the waterface, is very modern in that the downstream revetments have a function similar to the drainage blanket of more recent embankment dams. Without revetment even the small amount of seepage would soon waterlog the base of the dam and cause erosion. Finally, regarding the losses due to infiltration, one should note that silting in the reservoir itself would have the positive effect of sealing fissures that are natural to the basalt bedrock.

In discussing the phasing of the water systems it was pointed out that there was no stratigraphical proof in the strictest sense for the relation of the other two pools, P2 and P3. But their construction is very similar to that of pool P4, especially the dam about P3. The similarity in construction groups all three pools in phase 2. Of course one cannot be sure whether the apron of dam D1 also existed in the other pools, although that is likely. It is also uncertain whether the wall dividing pools P2 and P3 should be regarded as a spillway allowing the higher pool to fill before the lower and, as water is used, assuring that both reservoirs retain their separate portions. This arrangement would fit well into the compromise hypothesis discussed earlier since by the end of phase 2 gate LT1 is the most direct access to pool P2 from gate UT6 in the upper town. This was regarded as support for placing pool P2 towards the end of the

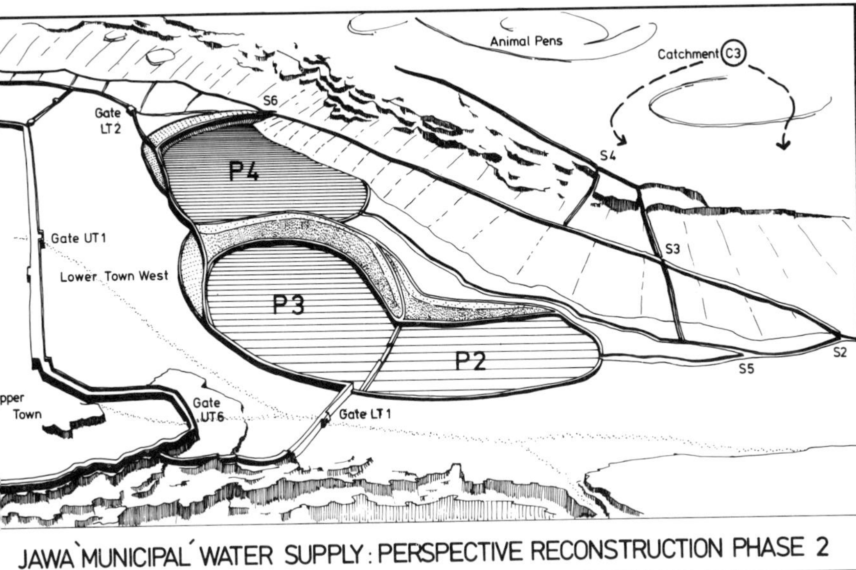

Fig. 83 Perspective reconstruction of area D

sequence. It should also be noted that if P2 is later than P3 a part of the latter's dam would be in this position.

With one exception the animal watering points were all built in the same way, essentially like the reservoirs next to the town. Location was dictated by topography (canal gradient) and natural features such as rock shelves and eroded basins. The three pools in line of system III are prime examples. There were probably natural faults in the basalt and tuff strata which were turned into water containers by filling gaps with masonry and earth. These artificial parts consist of two semi-concentric stone walls and the normal water-stopping ash and soil in between. Their cross-sections are very like that of dam D1 and their downstream profiles invariably the widest.

Only pool P1 is different. This is because a natural lava flow cave was used so that the pool acts rather like an aquifer near the surface of the basalt. Water was led into this underground cistern from the main gravity canal of system I (sluice gate S1). A small circular enclosure made of stone surrounds the mouth of the cave to keep animals away between planned watering times. The interior of the flow cave gave us proof of man's work in the form of crude stone

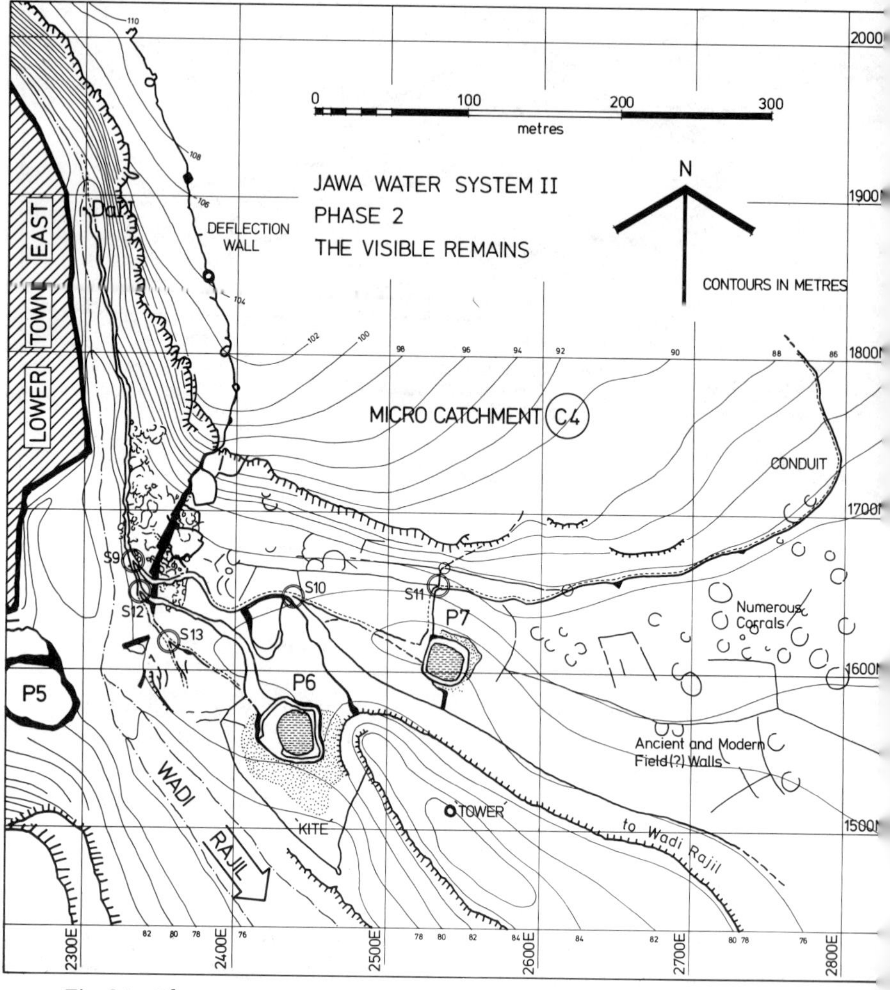

Fig. 84 Plan: system II

pilasters, apparently designed to reinforce the roof. Although the cave is quite small, compared to other pools, the considerably lower evaporation rate here under the shelter of rock increases the efficiency of the storage area. Pool P1 represents the most effective way to store water in a desert and the same method was used much later by the Romans and the Nabateans.

The various storage capacities of the ten pools at Jawa can be

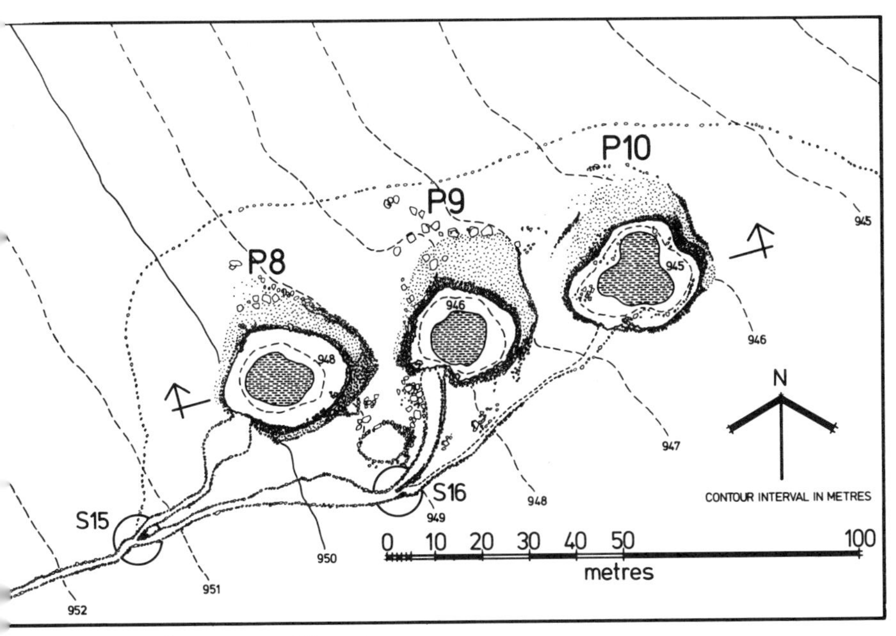

Fig. 85 Plan: system III pools

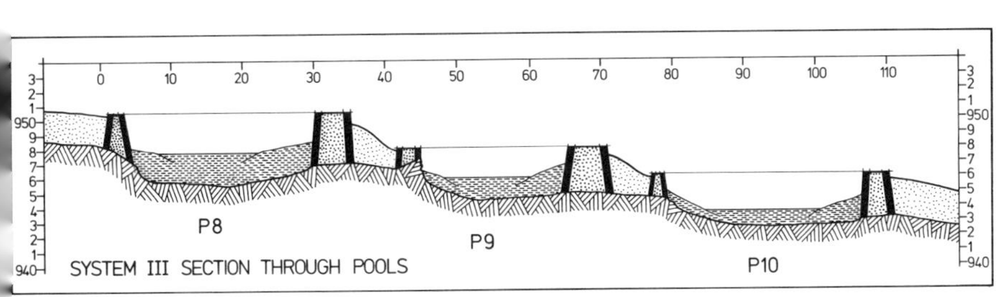

Fig. 86 Section: system III pools

calculated with some accuracy since the bedrock contours can be plotted. In general the lower volumes have been taken where there was any doubt. The animal watering points were measured less accurately with regard to their original depth, but reliable estimates are possible because of visible rock formations around them. A number of corrections must be applied to the absolute volumes. Various internal structures such as aprons, access steps (if they existed) and the probability of annual silting must be taken into account. However, these tend to be very small in relation to overall volume. Silting, for example, was very likely countered each year by using the mud from the pools for re-plastering houses, making bricks and perhaps even pottery. The soundings against the face of dam D2 showed that no appreciable amount of silt was left in the reservoir while it was in use during the urban time of Jawa. Of the corrections the most significant concerns the intended original maximum level of stored water in the pools and one might assume that this would be in the region of three-quarters of the total (absolute) volume.

Table 4 Storage areas: capacities

	Pool	Area (m^2)	V	VA	VA sub-total	VH	VH sub-total	System total
	01		500	500				
	02	2300	5994			4000		
I	03	5000	21,749		4300	15,500	42,000	46,300
	04	8400	30,821			22,500		
	05	1660	4989	3800				
II	06	1250	3771	2800	3800			3800
	07	530	1325	1000				
	08	380	950	700				
III	09	280	707	500	2000			2000
	10	430	1075	800				
TOTALS					10,100		42,000	52,100

V = uncorrected volume; VA = corrected volume (animals); VH = corrected volume (human). Volumes in cubic metres.

According to the discharge rate calculated for the gravity canals in system I (median rate of 0·6 cumecs) it can be estimated that the storage areas of the system might fill to capacity in just under twenty four hours. System II would require two hours and system III one. All of this of course depends entirely on steady flow of the correct magnitude in wadi Rajil combined with fortunate local rainfall and also does not take into account initial losses due to infiltration, seepage and evaporation when water first enters a dry system.

22

THE MICRO-CATCHMENTS

During the very first stages of settlement at Jawa locally generated run-off was the only source of water available for storage while a start was made building the shelters and water system. It has been argued that the technology implicit in the finished systems arrived at Jawa as a developed science. It follows therefore that the development seen here is a planned one and that very soon the major run-off source – wadi Rajil – was incorporated. This was done by extending the natural gully that fed the original pools next to the site. The new gravity canal was built along the western shore of wadi Rajil. Thus not only was the macro-catchment tapped but in the process of extending the canal northwards further smaller local catchments, or micro-catchments, were added too. These seem to be rather insignificant when compared with the mighty floods of wadi Rajil whose catchment area is 300 km^2, while that of the micro-catchments is hardly more than 1 km^2. Yet in calculating the water balance of Jawa it will be seen that these humble catchments at the side of wadi Rajil can be indispensable under certain conditions. Even if rainfall on Jebel Druze is relatively certain each year, it does not always come at the same time during the winter; nor do the effective storms, with regard to the macro-catchment, always oblige in their direction. Macro-catchment discharge at Jawa is therefore quite variable in time and volume. This is a vital factor in the overall water balance and the micro-catchments will be seen to serve as a kind of insurance policy. It is therefore essential to attempt to estimate their potential run-off yield in an average year.

The area and slope of the micro-catchments is known, but there is no empirical data regarding run-off yield per unit area. The soil cover is shallow and bedrock essentially impervious. Thus groundwater flow would be minimal and interflow negligible. Soil characteristics related to run-off imply a high rate and hence a high yield. The problem is how to enter such impressions into our hypothetical numerical discussion. There is also the more general question of whether our calculations have any bearing on the ancient environment of over 5000 years ago. It will have to be taken on faith to some extent that very little climatic change has taken place in the region since about 10,000 BC. If the water systems at Jawa can be shown to function under modern conditions there is a good chance that these calculations are realistic and at the very least reflect the principle and organization of Jawa's water budget. In any case it is known that the life-support systems must have functioned successfully – if only for a short time.

Referring yet again to the comparable data from the Negeb Desert, the table below sets out the calculations regarding the various micro-catchments at Jawa. Again the nomogram published by Evenari showing rainfall-run-off relations, including catchment factors such as size, slope and surface cover, has been used. The results are of course qualified by the normal variability of rainfall year to year, the path of storms, their intensity and duration, so that it is conceivable that in some years the yields would be below the averages listed here.

Table 5 Micro-catchments (mc): average annual run-off yield

	(mc)	area (ha)	slope (%)	run-off (m^3/ha)	yields (m^3/yr)	sub-total	storage (m^3)
	C1	38	2·5	200	7600		
I	C2	98	2·3	150	14,700	39,550	46,300
	C3	115	2·6	150	17,250		
II	C4	16	6·7	220	3520		3800
III	C5	143	6·0	160	22,880		2000
TOTALS					65,950		52,100

The figures show some relationship between the total storage capacity of the systems and run-off yield. This implies a very precarious balance if local catchments were the only source of water. The figures also demonstrate the possibility of surplus in some areas where storage requirements fall noticeably below the potential

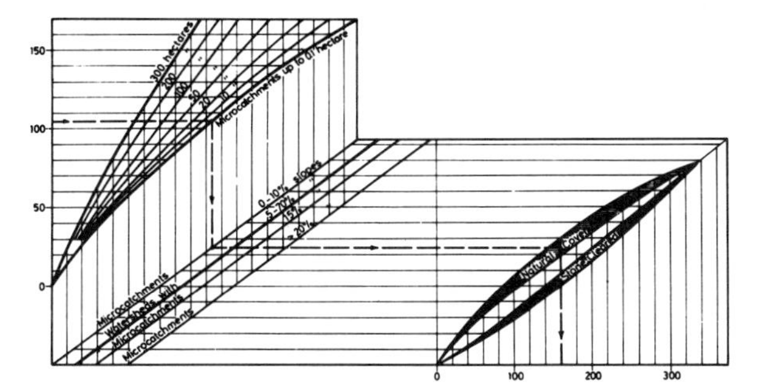

Fig. 87 Run-off rates at Avdat

yield, as in system III for example. Since we know that wadi Rajil contributed water, the other systems may also have occasionally carried a surplus. It is possible then to think in terms of irrigation of fields along the canals – as indicated in the schematic plan, figure 69, and as incorporated into the water-balance calculations below. However, when speaking about irrigation at Jawa one must be cautious. The only proof there is comes from small samples of carbonized seeds and some limited structural, locational features within the water systems.

23

AGRICULTURE

Preliminary study of carbonized botanical remains from various trenches has provided an initial idea about the Jawaites' diet as well as some evidence about land-use (see Appendix D). From the start of our work at the site the abundance of seed-processing tools suggested that agriculture played an important role in the food economy of Jawa. Irrigation was well established as an agricultural technique in Egypt and Syria/Mesopotamia long before the fourth

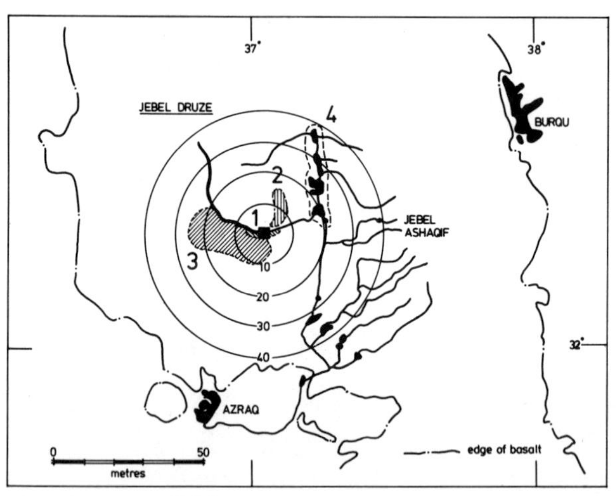

Fig. 88 Map: field systems

millennium; there is therefore a precedent, as well as the admittedly limited evidence from the site, to suspect that this method of growing plants requiring more water than the natural environment could provide was used at Jawa. The fields may be divided into four groups according to location, topography, distance from the town, soil cover and the kind of agriculture possible in them:

1 Fields associated structurally and topographically with the gravity canals of the water system: irrigation (?).
2 Wadis within 5 km of Jawa in which some signs of terracing for dry farming have been found.
3 Flat land to the west of the town site that can and does now support dry farming: up to 10 km from Jawa.
4 Mudflats that are inundated annually by run-off from Jebel Druze and its foothills: 20–40 km from Jawa.

The second and third areas need little further comment since they more or less explain themselves and since the evidence for ancient fields is almost certainly gone because of more recent activity. The botanical evidence is easily explained by the remaining areas.

The fields next to the gravity canals might have been irrigated by opening sluice gates once the storage areas were full. There is however very little structural proof of this, partly because of repeated reconstruction more recently, partly because the areas have been farmed again after Jawa, and partly because the structures that we seek to identify are just simple openings in an equally simple canal. There are two spots in the canal of system I where a gap was left on the downhill side which may once have been such a sluice gate. There are some fugitive signs of additional canals leading into the fields, but it is almost impossible to date them. If these were general features dating from the time of Jawa then local and regional run-off would have been used for irrigation – again, once the storage requirements of the town had been satisfied or stood a good chance of being replenished in due course. The fields numbered f1, f5, f6 and f7 (the majority) are nevertheless still questionable candidates for irrigation. Only fields f2 and f3 are definitely Jawaite installations and these too have been badly ploughed-over more recently: this much is obvious even in Poidebard's aerial photograph of Jawa which shows some very recent activity in field f2. Thus only field f3 provides any clear proof of having been irrigated as the terrace dams visible on the ground are definitely part of the ancient water system.

To be effective in terms of surface run-off the micro-catchment would have to be kept clear of vegetation as well as stones. The total area available for irrigation is about 19 hectares. For an equally rough estimate regarding the amount of water needed annually one

Table 6 Potentially Irrigable fields

System	Field	Area (hectares)	System (hectares)	mc Area	R
	f1	8·00			
I	f2	0·75	10·00	251	25·1
	f3	1·25			
II	f4	0·30	1·00	16	16·0
	f5	0·70			
III	f6	4·00	8·00	143	17·9
	f7	4·00			
TOTAL			19·00		

R = area of catchment/area of field : in the Negeb farms R ranged from 17 to 30.

can take the figure 3000 m^3 per hectare for barley which gives 57,000 m^3 of water per year: again, close to but in excess of the locally-generated run-off that could be expected in a good year. Wadi Rajil must have played an important part in any irrigation project; furthermore the interdependence of the two water sources is demonstrated.

The mudflats, especially those near Shubeiqa about 25–40 km away, are still used for agriculture. They not only receive flood water each year but some of this water is drained from the southern end of the chain of flats along the continuation of wadi Rajil. Drainage of mudflats is essential to soil fertility because without this periodic bath (leaching) certain salts carried down by the floods each year would stay in the soil as water evaporated, ultimately making the land useless for agriculture. It is a common sight in the basalt desert to find fertile and completely barren mudflats one beside the other. Khirbet Shubeiqa is the scene of some agricultural activity today, and it is the place to which the Jawaites have been thought of as coming from the north after their exodus. The beduin from near Jawa still use the area and will quite happily walk there and back in a day for a cucumber. If need be they will camp for a few nights during the harvest. Modern crops include barley, millet, many varieties of cucumber, tomatoes and melons. We have found hut-circles around the shores of the mudflats as well as the typical flint material suggesting that the Jawaites, their predecessors and also those who remained after the passing of Jawa all used Shubeiqa as their granary. The area of these potential fields is in excess of 1000 hectares and this alone, compared to the fields near Jawa, would make it the major staple crop producer during the fourth millennium.

Irrigable fields next to Jawa could have been flooded periodically during the winter months and chickpeas, peas and perhaps lentils would have flourished there. The larger areas, including the land to the west, would be used for staple crops such as barley and wheat. If as assumed the climate was roughly similar in the days of Jawa the agricultural calendar must have followed closely that of the modern bedouin. Adjustments might be made in efficiency terms – since Jawa was a highly-organized urban machine – but Jawaite paleo-science could not affect the weather, which imposed winter cultivation. The traditional beduin winter crop is barley, yielding between 0·4 and 0·6 tons per hectare in a good year. Avdat experimental farms produce 2·7 tons of barley per hectare (Evenari and Tadmore 1971) and Jawa might take a median place, producing about 2 tons per hectare in an average year. Sowing would take place only after the first heavy rains which could be as late as December. Thus in a dry year the harvest would come in May; in a wet year in March or April. In exceptionally wet years, when the rains came very early in the season and persisted throughout the winter, two crops may have been possible.

24

WATER BALANCE

We are now in a position to test the water systems of Jawa against the requirements of the town and also against the rigours of the climate in which they must have functioned successfully. To what extent the climate during the fourth millennium is comparable to that of today is not known absolutely. However, the relatively recent formation of the land, similar crop production, comparable wild flora and fauna, the presence before Jawa of Neolithic beduin rather than farmers, the absence of other towns in the region (and hence Jawa as an urban accident) and finally, the generally accepted lack of climatic change in the Near East since this rather recent chapter of prehistory all argue for a quite similar environment. Jawa's water systems worked, that much is known with certainty. If it can be shown that they would work under modern climatic conditions – especially if they *just* worked – conditions during the fourth millennium can only have been wetter. The ensuing calculations at the very least serve to enliven the albeit transient triumph of the Jawaites; and if the climate was like that of today the delicate balance of the life-support systems is dramatically illustrated.

First a realistic water consumption rate for the population of the town must be established. Various amazingly low rates have been quoted for human beings in the Near and Middle East, as was seen earlier in discussing the beginning of Jawa. From the Negeb comes the rate of 1·5 m^3 of water per person per year which is just 4·2 l per day. With this as a guide the daily water consumption for the average Jawa family of six might be as follows: 4·2 l of drinking water per person and an additional litre each for washing, while for cooking the average cooking pot (Appendix B) which holds about 14 l might be taken as the ration per family. The total daily consumption rate per family – a conservative estimate – would be

45·2 l. The daily rate for the whole of Jawa then, taking the maximum of the population range in phase 2 (3760–5066), is just over 38,000 l or 38 m^3 (or 13,920 m^3 per year; 1160 m^3 per month).

Animals consume the following amounts annually (Evenari 1971): sheep/goat 0·5 m^3, dog the same and donkey 1·0 m^3. On this basis one can suggest a maximum consumption rate of about 21 m^3 of water per day (or 7680 m^3 per year; 640 m^3 per month) to provision the following domestic animals at Jawa:

Table 7 Water consumption: animals

Animals	Number	m^3 per year	(% Appendix E)
sheep/goat	10,000	5000	(86·7)
cattle	800	2400 (?)	(08·5)
equids	200	200	(02·1)
dogs	160	80	—
		7680	

The total consumption rate towards the end of phase 2, animal and human, would be about 59 m^3 per day, 1800 m^3 per month and 21,600 m^3 per year.

What is important here is that the animal water supply in the specialized storage areas is finite, expecially during the dry season after May when no more rain can be expected until October. Depending upon the size of Jawa's own flocks, the amount of water available for barter with visiting beduin would in any case be limited. The presence of three large bodies of water at Jawa, the life support of the urban population, might easily have been a cause of crisis, as already seen before. In our experience of water shortage at Jawa the bedouin cared little about budgeting other people's water, particularly that of town dwellers. The nature of human contact, until now discussed mostly in terms of Jaywa, arises again. With undoubted extra-urban attention and pressure an agreement may have been made with the bedouin regarding water, perhaps in exchange for peace and breeding stock; an arrangement initially within the limits set by the town. This would apply primarily to the dry season when the pressure on both town and country was the greatest, but while settled bedouin within the walls of Jawa might at first co-operate others from without might not. They would not have cared much for, nor really have understood, the delicate life-support balance at the site since they themselves lived at basic subsistence level. They would avail themselves of any standard above that, if given leave and as long as it lasted. Their normal style of life was a part of their philosophy which included an inverted snobbery that takes the form, even today, of disdain and even loathing for the permanently-settled existence. Bedouin – and this would be as true of those who had settled at Jawa – can always return to the desert.

We come now to calculate this essentially unstable equilibrium at Jawa: the annual water balance in relation to modern climatic data. In what follows – hypotheses apart – it must be stressed that the data used are not entirely reliable because the recording periods are very short. This is particularly the case in the Jawa area and at the meteorological stations at Deir el-Kahf and H5. However, in general the climatic variables as they are shown to interrelate do conform to what is expected in an environment such as Jawa's and are

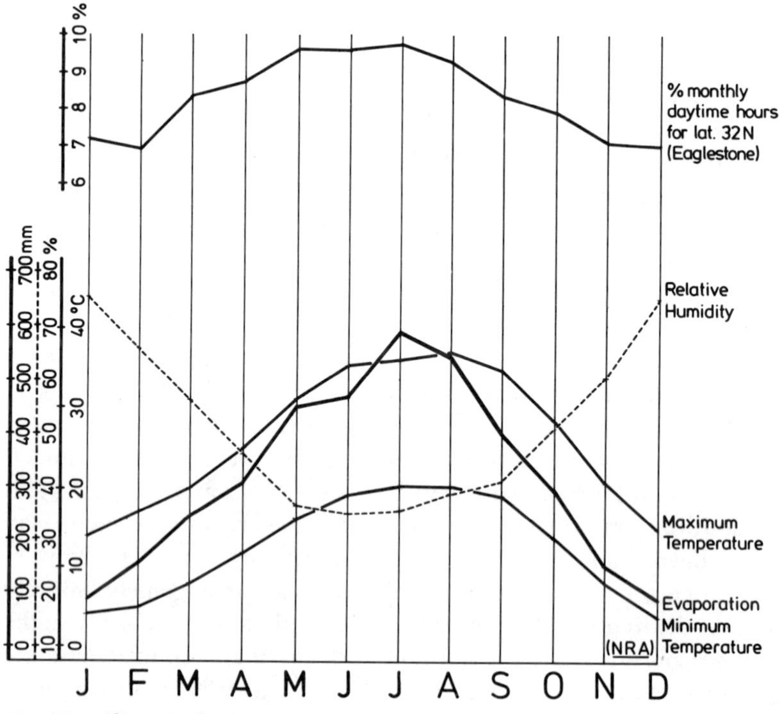

Fig. 89 Climatic data

furthermore comparable to areas where more reliable experimental data are available. This, after all, is all that is needed at this stage of the Jawa story, which is still preliminary in spite of the detail available.

On the 'negative' side of the hypothetical water balance losses due to free water evaporation are the most critical variable. In semi-arid regions such losses are substantial and figure 89 illustrates this very clearly. Evaporation is at its highest level when it is hottest, when the relative humidity is lowest and when solar radiation (expressed here in percentage monthly daylight hours) is most intense. However, the accurate measurement of evaporation on the one hand, and the use of such empirical data in an albeit simple environmental model here on the other hand, are problematical. The rates recorded at Deir el-Kahf, for example, are unrealistically high and had to be adjusted in these calculations. Application to Jawa's reservoirs and pools brings numerous factors into play whose effects cannot be detailed here; they are merely listed.

Besides meteorological factors such as solar radiation, air temperature, relative humidity and wind (strength, direction, turbulence and frequency of occurrence), geographical factors affect evaporation rates from open storage areas. They include elevation (affects atmospheric pressure and stability), water quality (salinity causes lower rates), depth of reservoirs (the shallower the higher the rate), size of water surface, land forms (shelter), surface water cover, temperature and so on.

The wind at Jawa is incessant throughout the year, coming mostly from the west and south. Incoming air therefore tends not only to be

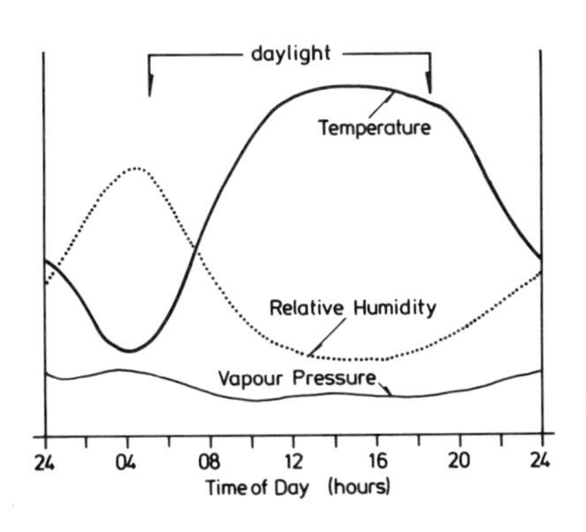

Fig. 90 Diurnal temperature, relative humidity and vapour pressure

dry – increasing potential evaporation rates – it also causes turbulence over the open reservoirs since they lie in an irregular wind tunnel made up of the basalt and tuff cliffs on either side. A high order of air turbulence in addition to normal convection removes vapour-laden air, reducing the saturated film over the reservoirs, and thus maintains a high transfer rate: a high rate of evaporation. However, certain factors tend to depress evaporation in a place like Jawa: an ever-present thin dust cover on the water surface and the quite low temperatures experienced at night, even in the summer months. The latter is demonstrated in figure 90, showing schematically diurnal interrelation between air temperature, relative humidity and their effect on vapour pressure. Because of the abrupt drop in air temperature each day a low basal water temperature is maintained as long as the water in the reservoirs is sufficiently deep. Obviously the time of crisis is mid- to late summer when water levels are necessarily low.

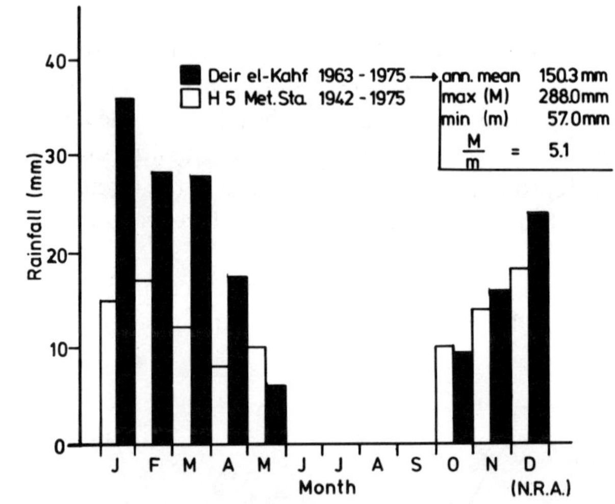

Fig. 91 Average monthly rainfall

In adapting the pan evaporation data from Deir el-Kahf to ancient Jawa the rates, as noted, have been adjusted downwards and then weighted according to probable percentage loss per month. Similarly the hypothetical run-off yields from Jawa's micro-catchments have been weighted monthly according to the probable percentage of

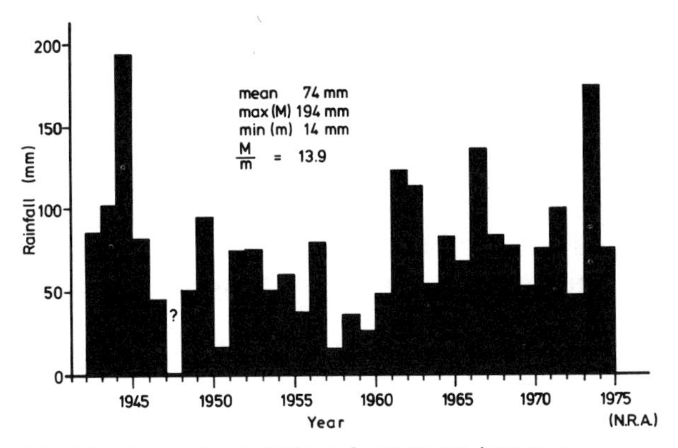

Fig. 92 Annual rainfall totals 1942–75 (H5 Met. Station)

precipitation per month. One must always remember, however, the hard environmental fact of Jawa: that high annual and monthly variations are the norm.

One of the most crucial factors on the 'positive' side of the water balance is the flood water in wadi Rajil. The mechanical problems of deflection have been discussed above. To this must now be added the fourth dimension, time: the time or times of peak flooding, some of which were covered earlier in the Negeb Desert examples and the hypothetical storm hydrograph in chapter 18. In the water balance calculations it is therefore necessary first of all to test the systems for peak flooding at various times during the winter and in the end settle for a probable time of flooding.

The water balance equation in general terms takes the form

$$\text{Inflow} = \text{Outflow} \pm \triangle\,\text{Storage}$$

in this case inflow being the total amount of water (input) from the macro- and micro-catchments (MC + mc) required to cover losses (E = evaporation, seepage, etc.), human and animal consumption (C) and irrigation needs (I), all together as output. These calculations were touched on in chapters 9 and 10 in discussing the Jawaites' arrival at the site. It was seen then that system I functioned successfully, relying for the most part on the local micro-catchments. Only when the consumption rate was raised to accommodate the bedouin settlers (Compromise 1) did the macro-catchment come into real consideration. The end-of-month storage curve and provisions for irrigation are shown in figure 94; but before considering this and the 'classical' water balance at Jawa in the fully developed stage of phase 2 it is necessary to examine the general character of this balance.

The water systems must be tested with regard to balancing input from the two catchments separately, together and at various times during the wet season against output. The following five 'situations' are examined, using the storage capacity of system I (maximum 42,000 m³), a micro-catchment yield of just under 40,000 m³ and a consumption rate of 1500 m³ per month (i.e.: Jawaites without bedouin):

(a) mc − (E + C): representing the water yield from the micro-catchments alone.
(b) MC^{no} − (E + C): representing the water yield from the macro-catchment (wadi Rajil) and a November flood, but no local water supplement.

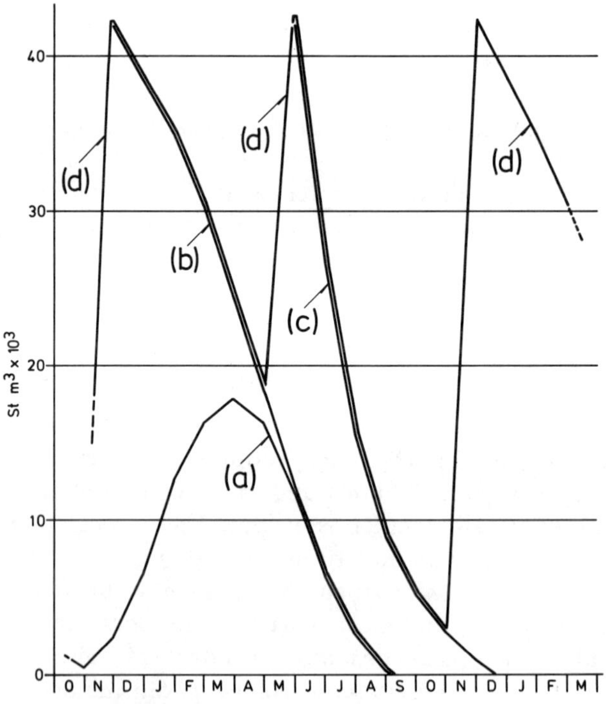

Fig. 93 Situations (a)–(d)

(c) $MC^{my} - (E + C)$: a flood in May and no local water supplement.
(d) $MC^{no} + MC^{my} - (E + C)$: no local water supplement, but large floods in both November and May (an 'overkill' situation).

With reference to figure 93 it can be seen that the micro-catchments alone (a) are not enough to support the Jawaites and their animals. Starting with the micro-catchment yield of just over 2000 m^3 in October the system is dry by early September. Similarly situations (b) and (c) show that single-flood input from the macro-catchment is insufficient: a November flood results in depletion about the same time as situation (a) and a flood in May, without the micro-catchment supplement, 'crashes' in the latter half of December. Only the 'overkill' situation (d) and a macro-catchment yield from two very large and widely separated floods allows the system to function: but at a most inefficient level since nearly 80 per cent of the water is lost to evaporation and other factors. Clearly therefore the answer is – and was, for that is what the Jawaites knew and did – to combine both water sources. And so it can be concluded that, small as they are, the micro-catchments at Jawa played a key role in the water balance and in the successful operation of the life-support systems. They ensured that the reservoirs and pools could be replenished throughout the wet season, independently of wadi Rajil; and since the macro-catchment could easily generate a surplus of water it may be proposed that the Jawaites made use of this for irrigation of fields along the gravity canals.

Returning then to the second year of the Jawaites after Compromise 1 and adjusting consumption of water by stages from 500 m^3 to 1410 m^3 per month (1160 m^3 for humans, 250 m^3 for animals, based on the storage capacity of system I, the stages reflecting population increases), keeping the storage areas at low level (2500 m^3) for November–January (the probable flooding in wadi Rajil coming after this), figure 94(6) illustrates the end-of-month storage curve. From the contribution of wadi Rajil in February a total of 30,000 m^3 is used for irrigation (barley requiring 3000 m^3 per hectare annually), just over 40,000 m^3 are used for human consumption and just over 9000 m^3 for irrigation. Pool P5 serves the animal population proportional to the human and any animals the settling beduoin might have brought (capacity 5000 m^3).

The water balance equation for Jawa takes the form:

$$mc_s + mc_i + MC_s + MC_i = E + C + I \pm \triangle St$$

and this can now be applied to the third year of the Jawaites, when the highest level of efficiency was attained – and, one imagines, maintained for a time.

End-of-month storage is shown in curve (7) of figure 94; the estimated amounts of water involved are presented in the table below for system I. Other animal watering points – systems II and III – provide water complying with a consumption rate of 250 m^3 per month for system II and 130 m^3 per month for system III (all systems, animal and human: C = 1800 m^3 per month). System II diverts *c.* 3000 m^3 for irrigation; system III *c.* 24,000 m^3. Total throughput for system I is:

70,000 m^3	animal and human use
30,000 m^3	irrigation
100,000 m^3	total

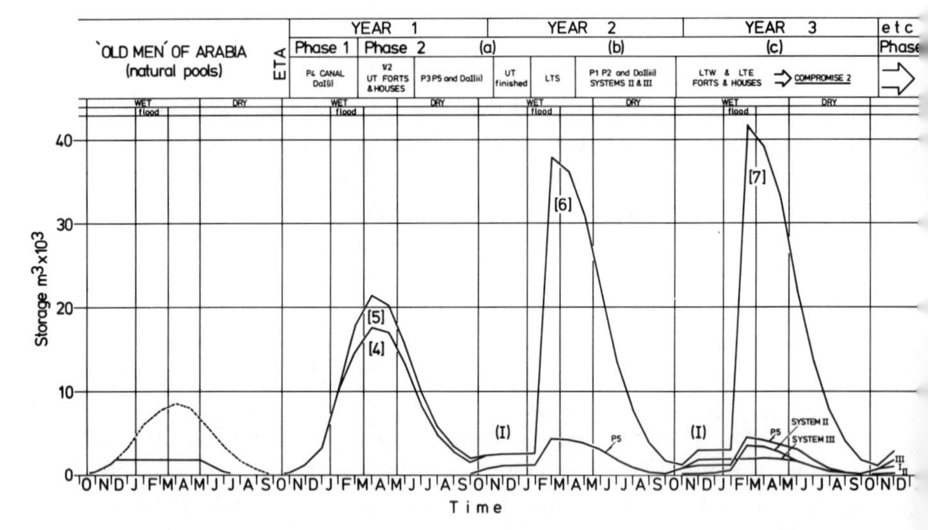

Fig. 94 Four years of Jawa

Table 8 Water Balance: System I (7)

			INPUT						OUTPUT				± △ STORAGE		
			mc_s		mc_i	MC_s		MC_i	E		C		I	St	
quarter	month	mc totals	H	A		H	A		H	A	H	A		H	A
I	N D J	18·24	05·81	01·39	11·04	—	—	—	00·85	00·29	03·48	00·75	11·04	03·00	01·15
II	F M A	17·56	09·04	01·80	06·72	44·43	03·82	12·24	19·54	02·20	03·48	00·75	18·96	33·44	03·83
III	J J	01·42	00·72	00·70	—	—	—	—	22·25	02·86	03·48	00·75	—	08·44	00·91
IV	A S O	02·33	01·06	01·27	—	—	—	—	04·50	00·63	03·48	00·75	—	01·52	00·80
		(39·55)	16·63	05·16	17·76	44·43	03·82	12·24	47·14	05·98	13·92	03·00	30·00		
					100·04						100·04				

(amounts in $m^3 \times 10^3$)

Comparative efficiency, regarding evaporation and other losses, during the 'classical' stage of the system (7) and the other 'situations' covered earlier,, is summarized below:

situation	E	$INPUT$	$\frac{E}{INPUT}$	$C(m^3)$
(a) mc − (E + C)	22·91	39·55	0·58	1500
(b) MC^{no} − (E + C)	28·29	42·00	0·67	1500
(c) MC^{my} − (E + C)	32·06	42·00	0·76	1500
(d) MC^{no} + MC^{my} − (E + C)	67·69	85·72	0·79	1500
(6) Phase 2 (beginning)	49·12	62·97	0·78	500–1410
(7) Phase 2 ('classical')	53·12	70·04	0·76	1410

amounts in $m^3 \times 10^3$

The admittedly rather hypothetical water balance demonstrates the interrelation between the two run-off systems that were joined together through Jawa's technological genius over 5000 years ago. It also underlines the importance of timing in such matters as peak floods with regard to human and animal water consumption as well as irrigation. The variability of most elements affecting the systems reflects the delicacy of balance which can properly be called an equilibrium between success and failure, maintained on the positive side by absolute control over water use and constant maintenance of the structures that made up the systems. Success depended as much on the peaceful resolution of differences that may have arisen between the Jawaites and their erstwhile labour force in the lower-town quarters of Jawa as survival depended on the weather. Failure was completely unacceptable. The systems were adjusted to accommodate the whims of nature but not necessarily those of men. If even one of the town's reservoirs was damaged during the dry season a disaster would ensue that would change the life in the town terminally.

The water balance described here carries the story through the third year of the Jawaites. Whether there was a fourth, a fifth or more depended on human nature.

VI

THE END

Detail from the Siege of Lachish (British Museum): urban refugees, once long ago nomads who conquered the land of Canaan and now faced exile – 597–582 BC

25

END OF COMPROMISE

Once again we have returned to the question of human relations, not only within the town and the potential confrontation between the upper and lower sectors but also beyond the walls of Jawa. The possibility of conflict arising out of demands that non-urban peoples might make on the life-support systems is a very real consideration.

Earlier the idea of co-operation was touched on. One peaceful form of this would be economic: trade between the settled and the mobile. It was seen that initially, even as far back as phase 1 and the beginning of phase 2, the 'ideal' was labour, seed stock and perhaps livestock in exchange for manufactured goods imported by the Jawaites. This changed almost as soon as it began and the interchange expanded to include settlement and a share in Jawa's urban technology. It became service for citizenship – up to the time of Compromise 2. Until then one must assume that the balance was kept and there was peaceful coexistence at the site, despite pressures and cultural as well as social differences.

The economy of the town may be described as essentially self-supporting – almost a closed system – at least in terms of basic needs to survive in the chosen environment. However, there were other, non-urban peoples in the desert who were the close cousins of the lower-town folk. What was the relationship with these desert people? Surely some form of trading would have occurred, as suggested above; and water would have been a dominant commodity during the dry seasons. But how would the town of Jawa have benefited? More labour? Perhaps at harvest time in the Shubeiqa mudflats? This seems unlikely in view of what happened earlier in Jawa's history and it seems also logical that the Jawaites would by now be limiting immigration to their town, if they had not been doing so all along. Trade in livestock is a possibility, but by this stage the town's own flocks would be thriving. Seed crops? The answer is the same: Jawa was established and self-supporting. Raw materials such as flint might have been bartered, but the Jawaites could easily have got these themselves; although that would require moving beyond the basalt which is a consideration of some interest since by doing so they would possibly have found the permanent water sources of el-Azraq and Burqu'. It seems then that there was almost nothing that the Jawaites could possibly want from the bedouin at large, the 'Old Men'of Arabia; nothing, that is, except to be left alone and in peace in their centralized, introverted and relatively comfortable town.

It is quite possible, and not without many recent parallels, that to be left alone was expensive. The bedouin's best-loved economic enterprise, then probably as more recently, was raiding – each others' camps, but preferably fat towns. Failing that they would occasionally resort to extortion. Thus one can complicate the ancient economic structure just a little more by including this element and arrive at a system that may still have been balanced but was definitely under some stress.

Let us assume that the weather was the least worry; that the life-support systems functioned reliably. And let us for the last time look again at the human situation at Jawa: the collective psychology of the fourth millennium desert town and the effect of population pressure.

At the beginning of phase 2 the population of the upper town was estimated between 1520 and 2280 men, women and children. The lower town began with about 1272–1908. This was Compromise 1: the southern lower town plus a part of the north-west sector. The situation changed quite rapidly and population of the lower quarters increased as is shown by the physical growth of these areas. The new figures any fourth millennium Jawaite anti-immigration alarmist might quote now would be between 1858 and 2786 (Compromise 2). A good chance exists that the Jawaites were already outnumbered, as well as being surrounded and cut off from their own water supply. This is so even allowing for some natural growth in the upper town 'founder' population. But it is suggested here that the time span of this urban drama was so short that population changes must be seen not as the result of procreation but immigration. Compromise 2 and phase 2 generally is the development stage described in such detail above and rightly called the highest ever reached at Jawa: a great urban achievement under very trying conditions, a triumph over inherent instability and therefore a momentary victory over chaos. In short, 'classical' Jawa was a town with built-in obsolescence, a successful failure.

By the end of phase 2 – and it could not have been very long since the start of Jawa, less perhaps than one generation – a disaster befell the town.

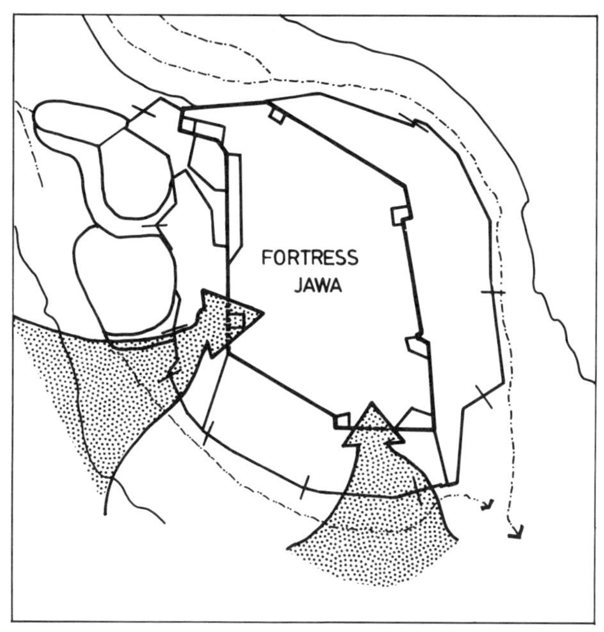

Fig. 95 Plan: end of compromise

Volcanic ash destroyed Pompeii and Herculaneum, sand dunes have obliterated Saharan and Arabian cities, and tidal waves, flood, earthquakes and a long list of natural upheavals have demonstrated the frailty of man's works on this planet. But this is not so long a list as the examples of man's self-destruction, of his almost pathological tendency towards genetic suicide. On two counts one could blame human failure for the end of Jawa: man's bad record afterwards and population pressure, an example perhaps of technology having exceeded wisdom. At the same time one cannot exclude from this somewhat pessimistic 'catastrophe theory' the other potential disasters that were realities of the time and place and probably helped to create the urban neurosis Jawa's citizens suffered. It has to be accepted that the water systems were not infallible; nor can one deny the chances of crop failure or other natural causes. But these cannot be measured like the inexorable growth of Jawa and the changes that would certainly have occurred in the town's collective psychology. It is obvious even now, much over 5000 years later, that a physical struggle took place at Jawa for control of the upper town. The strong and imposing black basalt fortifications there were partly destroyed in two key areas.

In area F where we excavated the town wall the fortifications suffered an inward collapse at the end of phase 2; but this was not just an isolated incident. An entire section of the upper-town curtain over 50 m long was torn down and the same happened in similar magnitude near the south-eastern corner. The repaired breaches are visible today and that means the struggle was not terminal so far as settled life was concerned. But certainly there was a mighty attack aimed at the administrative intellectual core of Jawa. Furthermore, other sectors of the town were similarly affected, suggesting a truly widespread disaster. The outer fortifications in area LF collapsed inwards and were later rebuilt. The new post-war stage at Jawa is phase 3, the last of most ancient Jawa. The disaster marks the end of phase 2 and the end of an ideal in more ways than one. To describe the final days of phase 2 one must extrapolate backwards from the next stage and thus describe a further population rise: for during phase 3 the water systems dealing with the municipal supply were altered and adjusted to a considerably higher potential storage capacity.

The expansion in area LF during phase 2 may be paralleled elsewhere along the southern flank of the town. It should be noted that with Compromise 2 the physical limits of growth had been reached here. There are two bulges between gates LT5 and LT4 and between gates LT4 and LT3. Undoubtedly this is proof of population pressure from within Jawa's lower quarters: and of a loss of control, since the orderly line of the fortifications is finally disrupted. There are further signs of this. Hut-circles appeared beyond the walls of the lower town, along the revetted fields beside pool P5 and even across wadi Rajil above sluice gates S9 and S12. This not only implies internal pressure, it points to the probability of further immigration in spite of whatever limiting policy the Jawaites tried to impose. Adding all of these areas together, including the over-building of the ring-roads in phase 3, one arrives at a new population range for the lower quarters towards the end of phase 2 and the beginning of phase 3 of between 2500 and 3700. The probability of bedouin numerical superiority is now quite definite.

With the idea of squatters and the heavy reconstruction programme that followed the attack on the upper town it is not impossible to think in terms of a large number of bedouin – not a new wave of urban folk from without the Black Desert – coming to Jawa in the dry season and wanting water. How many there may have been can be determined roughly from the projected new water supply of phase 3 (next chapter) that was planned to support a maximum in excess of the joint Jawaite-settled bedouin population of 5066.

Some time in the summer of a year – fourth, fifth of the Jawaites? – in the later part of the fourth millennium a mass of desert folk would have stopped on the high ground west of Jawa and negotiations would have broken down. The heat without enough water would have been as vexing as the sight of Jawa's apparently inaccessible pools. An attack was launched on the town, the battle array being as follows: manning the walls of Jawa about 760 Jawaite fighting men and about 1244 settled bedouin from the lower quarters, 2004 troops in all, facing (let us say) 1600 or more desperate, highly-motivated and, for once, unified desert fighters. A number of points in the lower defences were stormed simultaneously and for a while the Jawaite coalition held the walls and gates. One or several points were finally overrun and the relatively few Jawaite commanders killed. Defending troops now stood face to face with their cousins and found themselves fighting on the side of their recent employers, their leaders: the Jawaites who, like the forbidden upper town, no longer seemed so invincible. At the breaches in the lower town the suburban troops turned about and joined with their desert brothers and the mutiny spread and turned into a combined attack on the gates and walls of the upper town. Walls were breached and as the heavy black building blocks crashed down the slopes the upper enclave of the Jawaites lay open before them. The fighting odds had changed to about 30–7 – in favour of the desert.

26

NEW JAWA: A FAILURE

One does not know of course what happened when the fighting finally stopped on that summer's day over 5000 years ago: whether the killing went on into the night and eventually all of the Jawaites were put to death, or whether at some hopeless point they surrendered. It cannot be known whether, having lost the fight, they now abandoned their town and embarked on yet another exodus, or whether they made terms with the victors and stayed, although no longer as the ruling class. All that is known is that after the battle Jawa continued to be a town of a kind.

The water systems were rebuilt along rather different lines and one can deduce a final chapter in the history of the site because the new scheme was in the end abandoned as a failure. It is possible that whoever of the Jawaites remained now quit the town and the Black Desert, before this final phase. Where they went is another question, alluded to at the very beginning of this book. When they went is not so important since the time scale for the whole of the urban Jawa phenomenon is so short. In any case, it will be seen shortly that a second water crisis was in the offing.

Whatever the fate of the Jawaites may have been, and even if some of them remained after the battle, living among a triumphant and liberated people they had once ruled, it is very clear that the new order did not heed the old. Town planning as a conscious choice was at an end. Streets were overbuilt in the upper town and the ring-road along the fortifications was abandoned. The very fortifications now served as the foundations of new domestic architecture just as the walls of innumerable ancient cities throughout the world would become slums once they no longer served a military purpose. At Jawa during this final phase fortifications seemed no longer part of the urban matrix and only the

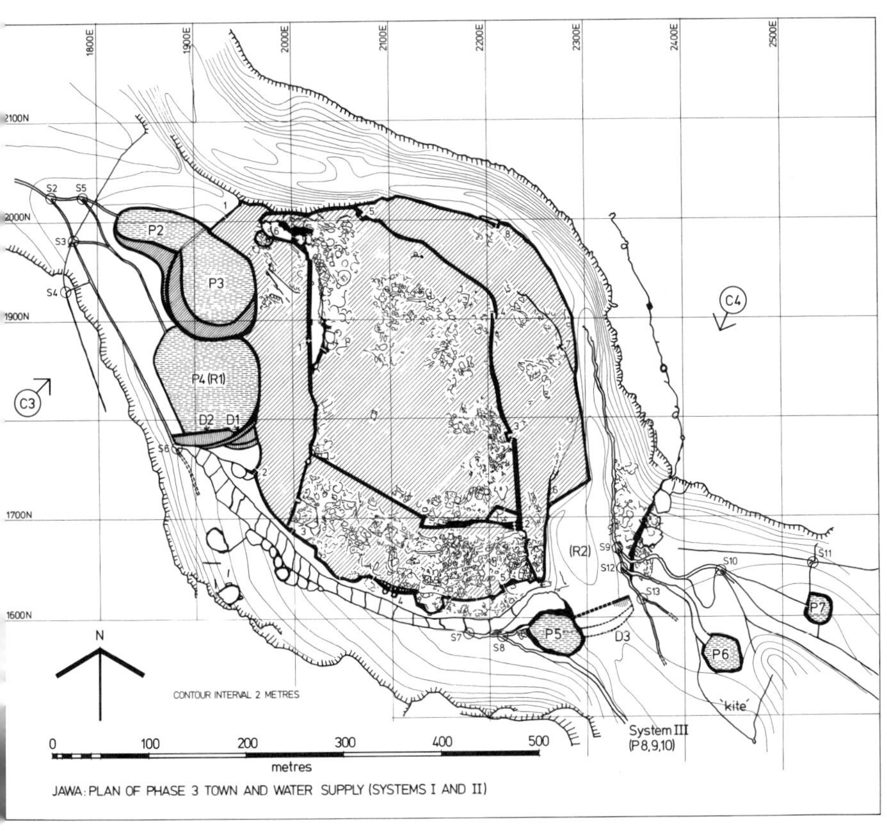

Fig. 96 Plan: Jawa in phase 3

water systems – out of necessity – received any communal and concerted attention. Water, after all, had been the cause of war and its prize. But could the people of the lower quarters and their brothers from the Black Desert maintain this prize? Had they learned enough from their masters? And, one may well ask, would we today who are 'civilized' and exposed to science be able to guide Jawa through the summer and the winter and reinstate the complex water supply of the Jawaites as well as – probably the most vital consideration – guiding the new lords of Jawa, persuading them to practise restraint, order and conservation?

The new plan for the water supply was a bad copy of the original. Although the new dam across the valley west of Jawa impressed all

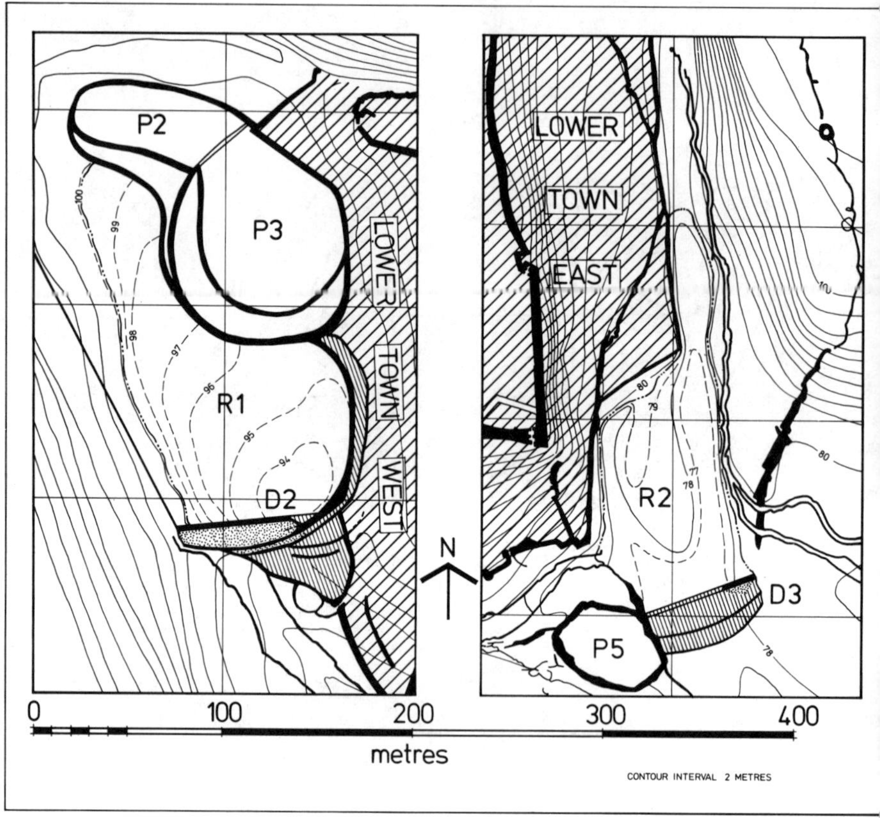

Fig. 97 Plan: water storage as planned

who saw it thousands of years later it was only an imitation of the fortifications: big, black, straight and virtually useless.

Both this straight dam (D2) and the very obviously failed one in wadi Rajil (D3) are placed in this phase. It can be seen how effective the latter was because only the sad stump has survived against the eastern shore of the wadi. The dam was probably amputated by the first winter flood. Also it is possible that like its big brother to the west it was never completed. However, since the planned storage capacity of the new system has been used to measure population pressure on Jawa a reconstruction should be attempted. After all, it is not so common in preliterate archaeology to be able to report not only what was built but also what was planned.

The absolute maximum level of the shoreline that the new lake behind dam D3 could have reached would be the 79·5 m contour line because the gravity canal leading to system II would be submerged by a larger reservoir. The new storage area is therefore a very small gain indeed for an engineering adventure that would exceed anything ever contemplated at Jawa before. Reservoir R2 was to have had a surface area of about 8500 m^2, an average depth of 1·6 m and a total volume in the region of 13,000 m^3 with a corrected storage capacity of about 10,000 m^3 at the very most. If the new dam had been completed by October the reservoir could have been filled partly with surplus water from system I along the by-pass canal west of dam D2 and maybe in this lay the genesis of Neo-jawaite planning. Someone had perhaps watched water returning to wadi Rajil and thought it very clever if that too could be harnessed. The problem, as is only too well known, is not this tiny trickle lost but the flood soon to come from Jebel Druze. It was madness – or just blindness? – to think that any dam, however well designed and strong, could withstand that for long.

The situation was less drastic in the area of pool P4. This was now to become the peoples' municipal reservoir, R1. The planned height of the new dam – and one must assume that a bigger and better lake was the intention – may be taken as the 100 m contour line because that is the highest surviving point today: in relative height just below the spillway and canal leading to pool P5. Our excavations have shown that this new dam was never finished. There were no traces of it beyond the presently visible end. One could argue that a flood washed out that part, but there are two strong objections to this. First, a flood sufficiently strong to bring down a virtual monolithic redundancy like this could not be generated by precipitation on the micro-catchments in the west. One cannot seriously suggest that deflected water from wadi Rajil would do such damage because the gravity canal would burst long before this and disperse the water. Only manual destruction could have cut the dam and there is no compelling reason to suspect that. Secondly, the unfinished state of the dam can be very simply verified by measuring absolute heights along the 100 m contour, the shore of the proposed reservoir. This shows that if the lake were ever filled a rather substantial part of the western lower town would be innundated. In order to contain the lake and protect the town a heavily revetted dam about the same size as D2 would be required; and this was never built.

However, even if it did not work out as planned the calculations should be taken on to their unhappy conclusion. The new reservoir was to have a surface area of about 12,500 m², an average depth of 3·6 m and an uncorrected volume of *c*. 44,000 m³. Corrections would reduce this to about 31,000 m³ which is an improvement of just over 9000 m³ over pool P4 of the previous phase. The total projected storage capacity of the new scheme is summarized here and represents the municipal supply; animal watering points, poignantly, remain the same and show once more the lack of planning ability of the Neo-Jawaites.

	m³
Reservoir R1	31,000
Pools P2 and P3	15,500
	4,000
Reservoir R2	10,000
	60,500

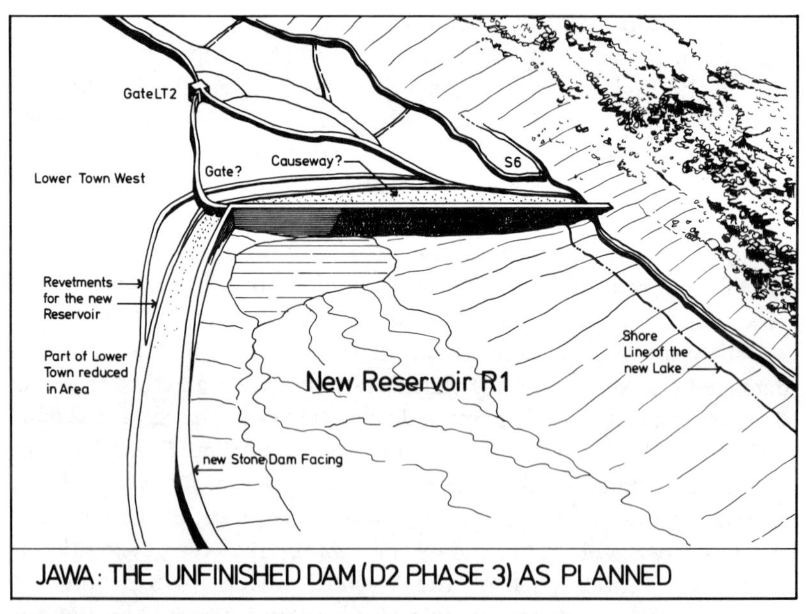

Fig. 98 Perspective reconstruction of the new scheme

The new storage capacity is an increase of about 45 per cent over phase 2 but does not, upon closer examination, allow a commensurate rise in the consumption rate, whatever the Neo-Jawaites might have hoped, to accommodate their desert brothers. Figure 100 illustrates the end-of-month storage for a rate of 2000 m^3 per month which reflects a new potential population figure of over 5600 according to census calculations: an increase of less than 600. These calculations, moreover, do not allow for irrigation based on water from the micro-catchments, nor does the new scheme increase the capacity of the animal watering points, as noted above. Reservoir R2 (capacity 10,000 m^3) does not add much either, despite the engineering feat it implies. In figure 100 normal losses without any water consumption at all produce a very unpromising storage curve and even a low consumption rate (250 m^3 per month) results in failure at the beginning of November because the new reservoir could only be fed effectively by winter floods in wadi Rajil. Even if the new scheme had succeeded an overestimation of stored water resources would have led to conflict, ironically this time among the new bedouin lords of Jawa. By November – perhaps just a mere sixteen months after the water war – pressure on the water systems would become unbearable. Neo-Jawa seems to have been a very poorly planned project that sought to accommodate more people

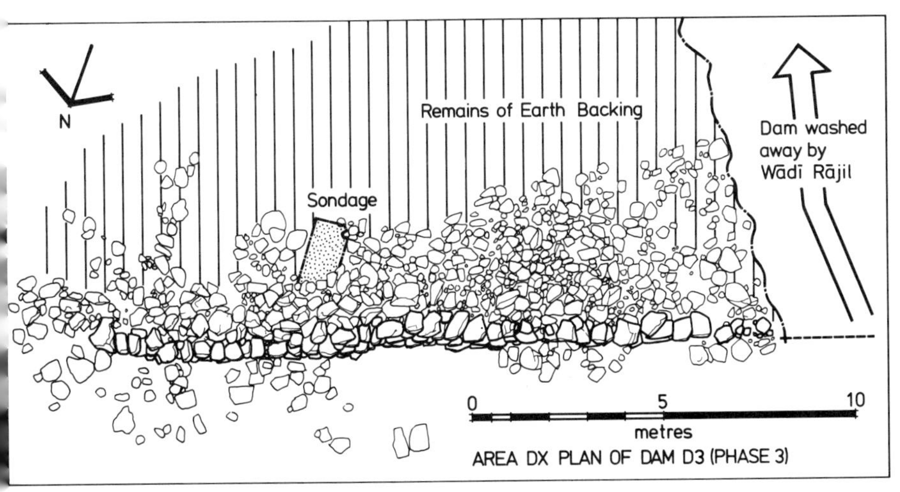

Fig. 99 Plan: the failed dam in wadi Rajil

than the environment could support – even assuming that the newcomers were meant to displace the original Jawaites. In addition to this the new improved constructions that were to achieve all of this had to function properly – which they did not.

The whole grand scheme failed and tons of soil and stone shifted and disintegrated in wadi Rajil, probably some time in January or February. Dam D2 was never finished. Perhaps this was because the rains came, the water began to rise in the unfinished reservoir and the red-faced engineers realized that there was not enough time in which to shore up the western lower town. The gap between the new dam and the town was hurriedly plugged and it is possible that some water overflowed at this point causing the earth fan still visible today. If all of this happened the new scheme would simply have been the same as that of the previous phase: and ironically the reservoir would hold less water because of the bulk of the new dam.

Without the Jawaites (the best explanation) and suffering from a fourth millennium 'brain drain' the new 'Old Men' of Arabia might have struggled along at the site for a season or more, slowly losing interest in maintaining the complicated systems that in any case were the legacy of an alien people. The technology implied by these systems, as noted earlier in this story, was not that strange once it was explained and demonstrated. The elements that made the whole were, in themselves, simple and close to aboriginal desert hydro-technology of a kind that the 'Old Men' had known well before Jawa. The difference lay in scope and organization: in a form of civilization. And it seems that the 'Old Men' either never achieved this or, if they did achieve it, did not wish to retain it. So people would gradually drift away from Jawa, to the more normal life they had known previously, and Jawa would become under-populated (irony of ironies) and die a slow but inevitable death of old age as the various life-support systems functioned less and less efficiently until they finally stopped for ever.

The fate of the Jawaites apart, the settled nomads would now once more become bedouin, completing one circle of the demographic pattern common to the region and the Near East. They might have gone back to the desert to inspire settlement types such as the ones described by Maitland, the so-called hill forts, if this is where these places belong chronologically; or they might just have disappeared as far as the physical archaeological record is concerned, not to be seen again until the words of their heirs were carved in Safaitic many hundreds of generations later.

Jawa became a ghost town and wadi Rajil flowed each winter as it had done before Jawa, and bedouin came and went and sat under the

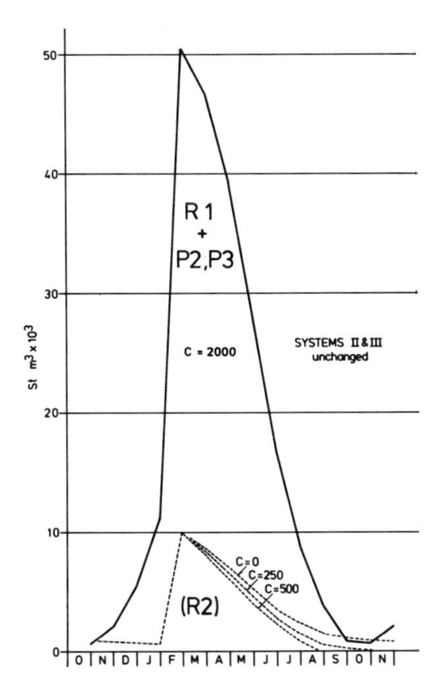

Fig. 100 Phase 3 as planned

cliffs opposite the ruins above the dams and probably had not a clue of what had passed. Nor cared. And so ended the first and really quite accidental and even unnecessary urban experiment in the Black Desert to occur between the volcanic formation of the land and the Hellenization of the Nabateans. The non-urban human record traced here from about 8000 BC up to the time of Jawa continued, as did the great demographic cycle in the rest of the ancient world. The Jawaites might have had a part in the urbanization of Palestine; after the brilliant but needless false start at Jawa they might at long last have found peace in a prehistorical Promised Land.

The cycle of human aspirations continues to the present day in these lands and the Jawaites are merely one very humble example of the earliest known exodus of a 'civilized' people cast into the wilderness. Jawa is among the earliest records of man's genius and frailty; above all of his underdeveloped talent for peaceful coexistence.

Wall-painting from a third millennium Egyptian tomb at Deshasheh: siege of a Palestinian (?) fortified town

APPENDICES

A) *Stratigraphy and phasing*

B) *Artefacts and chronology*

C) *Qa'a Mejalla survey (1979)*
by A. V. G. Betts

D) *Plant remains*
by G. Willcox

E) *Animal remains*
by I. Köhler

F) *Material requirements and labour-time*

G) *Notes on some Safaitic inscriptions*
by M. A. C. Macdonald

APPENDIX A

STRATIGRAPHY AND PHASING

The stratigraphy of Jawa is quite clear; the task of recognizing distinct 'phases' of occupation is therefore an easy one. What is more – and this is rare in the archaeology of extensive sites – widely-spaced trenches produce remarkably similar evidence so that all of Jawa can be understood in terms of one beginning, a development and a clear end. As proof of this three key areas of the town can be cited: Area C Trench II against the Middle Bronze Age Citadel at the summit of the upper town, Area F Trench II against the fortifications of the upper town and Area LF Trench II against the fortifications of the southern lower town. The overall results are summarized in table A1.

Area C Trench II (figure A1)

(1) Ashes, charcoal and sherds lie on virgin soil and bedrock just a few centimetres below the footings of the Middle Bronze Age Citadel. There are two sub-phases: a series of ashy occupation levels and a small circular bin.

(2) The bin of the previous phase is sealed by ashy soil and a house with two rounded storage bins is built. There is quite a substantial series of floors and repeated rebuilding of the bins. Phase 2 ends in flames and ashes.

(3) The door in the earlier house is closed and several floors run against the blocking stones suggesting that the house was reoccupied for a short time after the fire. Then follow the normal signs of abandonment and a very long period (over one millennium) of water- and wind-deposited soils.

(4) The Middle Bronze Age Citadel is built and two separate sub-phases can be recognized in the superimposed thresholds of the main

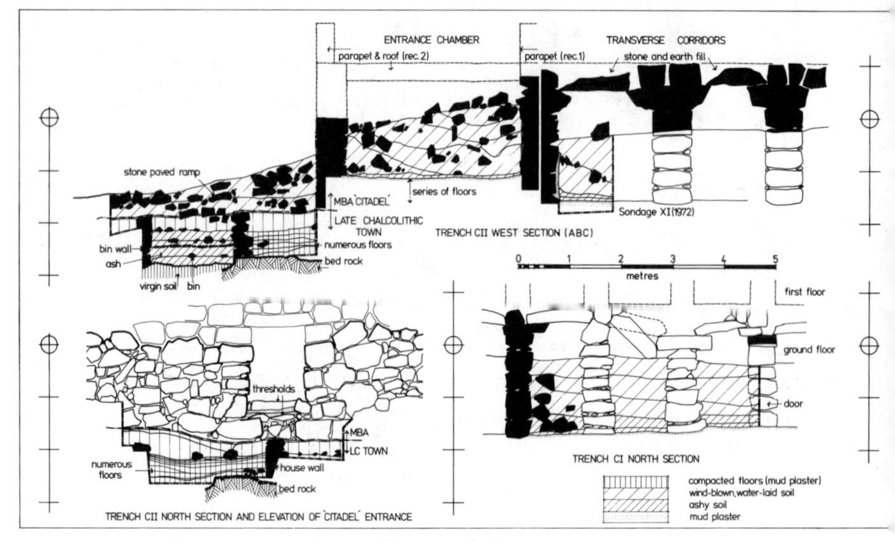

Fig. A1 Section: area C trenches I and II

entrance. This is followed by the final abandonment of Jawa in terms of permanent human occupation.

Area F Trench II (figure A2)

(1) There is an accumulation of ashy soil on bedrock and virgin soil (Section BB and CC) and this occupational debris, the earliest here, passes *under* the heavy upper-town fortifications. This is even clearer in Trench I some metres to the east where the ashy layers are over 50 cm deep. Jawa was therefore settled before the construction of the fortifications.

(2) The upper fortifications are built. Two sub-phases can be noted: the first is the beginning of military architecture and the deliberate siting of a street along the interior of the curtain; the second is represented by the so-called industrial installations. A 'kiln' is built and repeatedly fired, up to the end of the phase. At that point in Jawa-time a new clay lining is added to the 'kiln' but the installation is never used. The collapse of the unfired 'kiln' and a partial inward collapse of the fortifications separates phases (2) and (3). This event signals a crisis at Jawa even more than the fire in Area C which on its own could merely be a local accident.

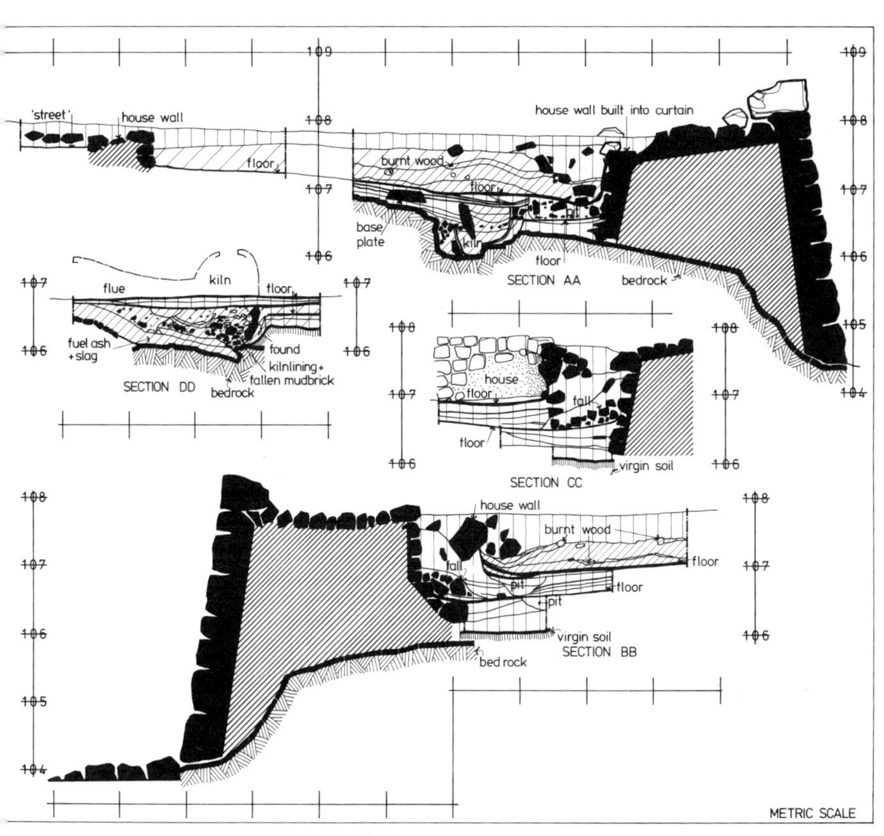

Fig. A2 Section: area F trench II

(3) Something quite new occurs, both structurally and in terms of planning. A typical domestic dwelling is built into and even onto the partly collapsed town wall. The floors, pits and a roof support baseplate are quite clear in the section. Phase (3) ends with a serious conflagration that has left charred roof beams, wooden roof supports, burnt mudbricks and general debris, all of which slowly moulded itself into the grey-black mass visible today as a millennium of rain and wind ruled Jawa.

(4) There is no orthodox stratigraphical link with the end of the last phase as there is in Area C, but at the northern end of Area F lies one of the clearly different structures belonging to the Middle Bronze Age Citadel complex.

Area LF Trench II (figure A3)

(1) Ashy layers pass *under* the earliest fortifications (Section AA: wall B) as in Area F. However, they do not appear beyond the line of the fortifications, suggesting that this is also the southern limit of the first domestic occupation of Jawa.

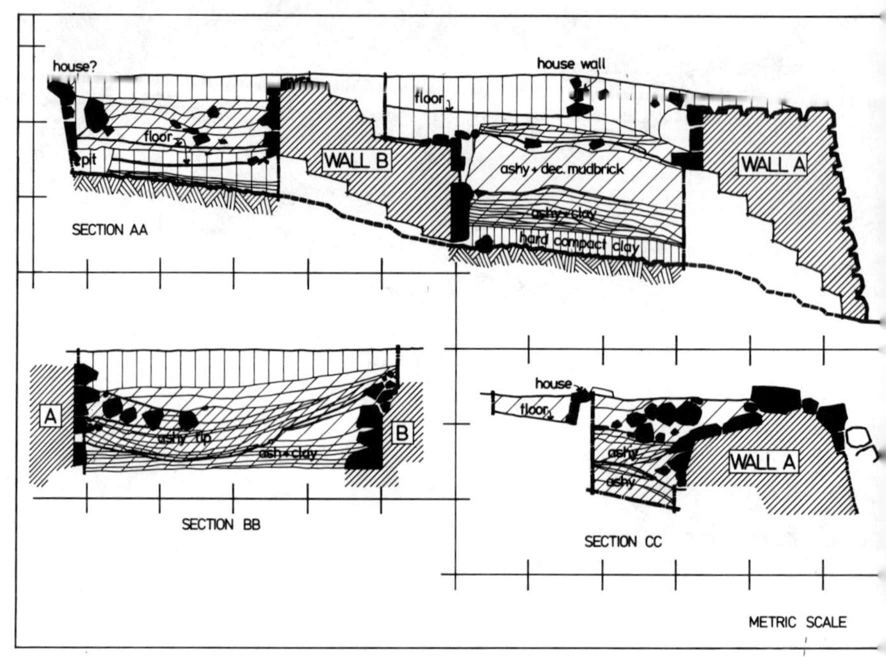

Fig. A3 Section: area LF trench II

(2) Wall B is built as the first fortification system of the southern lower town. Farther to the east gate LT4 is installed (Trench IV). The construction of a house follows (wall C) with a series of floors and pits. At the same time debris builds up outside the fortifications. There are signs of deterioration in wall B and this is followed by the next major sub-phase, the construction of a new line of fortifications (wall A). The lower town is expanding. A postern is set into this new line to the west of the trench and the space between walls B and C is filled in, while ashes and decomposed mudbricks accrue against the outer face of wall B. This could be a deliberate fill between the two lines of fortifications. A street may have existed here along the inner

face of the curtain, just as in the upper town in Area F. There are signs of erosion tipping towards the postern. Finally, as in Area F, the outer fortifications suffer an inward collapse and this is followed by phase (3). The evidence once again points to a crisis.

(3) Again as in Area F, a house is constructed against and over the fortifications. A short time later the area is abandoned. There are of course no traces of phase (4) in the suburban parts of Jawa.

Table A1
Summary of phases

The 'Old Men' of Arabia
(Late Neolithic bedouin)

ase	*Area C*	*Area F*	*Area Lf*	*Comments*	*Water systems*
	ashes bin	ashes	ashes	first settlement: open and unfortified; first domestic architecture	system I: P4, canal and DaI(i)
	house fire	curtain 'kiln' collapse	ashes walls B and C floors wall A fill collapse	upper fortifications begun; *Compromise 1*; lower town (south) and eventually west and east quarters built; *Compromise 2* and lower town expanding	(a) P3, P5 and Dal(ii) (b) P1, P2 and Dal(iii) (c) systems II and III
			Crisis: breaches in upper and lower fortifications		
	door shut fire	houses on curtain —	houses on curtain —	upper and lower towns reoccupied, but much more densely built up; then abandonment	new water scheme planned, revised and abandoned
			A millennium of wind and rain		
	threshold 1 threshold 2	—	—	Citadel and outbuildings built, used for a time and then abandoned	parts of the water systems re-used
			More wind and rain *Bedouin* *Safaitic inscriptions at Jawa* *etc.*		

APPENDIX B

ARTEFACTS AND CHRONOLOGY

1) *Pottery*

A correlation of the Jawa pottery with dated assemblages in the Near East is given in table B1.

Pottery from the Middle Bronze Age Citadel complex consists of vessels normally associated with domestic life. There are crude open cooking pots with ledge handles near the rim (figure B1.1), finer wheel-made decorated cooking pots (figure B1.2) and heavy storage jars (B1.5, .10 and .17) as well as finely-made serving bowls (figure B1.3). These would meet all the needs of processed foods which would be served up in the many smaller containers shown in figure B1: .4, .13–.16. There are also a few of the typical Middle Bronze Age II burnished 'button-based' juglets (figure B1.8 and .9) used to decant liquids from larger storage vessels. For parallels see Kenyon (1979 and 1960–5) and Amiran (1969). The complex was built and occupied about 1900 BC.

A functionally similar assemblage was found in the urban Jawa of the late fourth millennium BC. It differed mainly in being almost entirely hand-made. Containers shown in figure B2.1, .2 and .4 were used for storage: .1 and .4 for grain, .2 for water. Types B2.3 and .5 – both very common at Jawa – were cooking pots. Water and liquid foods were probably heated by inserting hot stones and for that reason practically all of the hole-mouth jars (B2.5) were found with their bases crushed beyond conservation. Many of the vessels were decorated with impressed, incised, gouged or scored designs that were added before firing. Several graffiti occur. They were cut after firing (B5.13 and .14). The range of designs in one of the typical cooking pots is shown in figure B5. Painted decoration is rare.

A second group of vessels consists of small pots that were used for conveying liquids (perhaps processed dairy products) as well as for

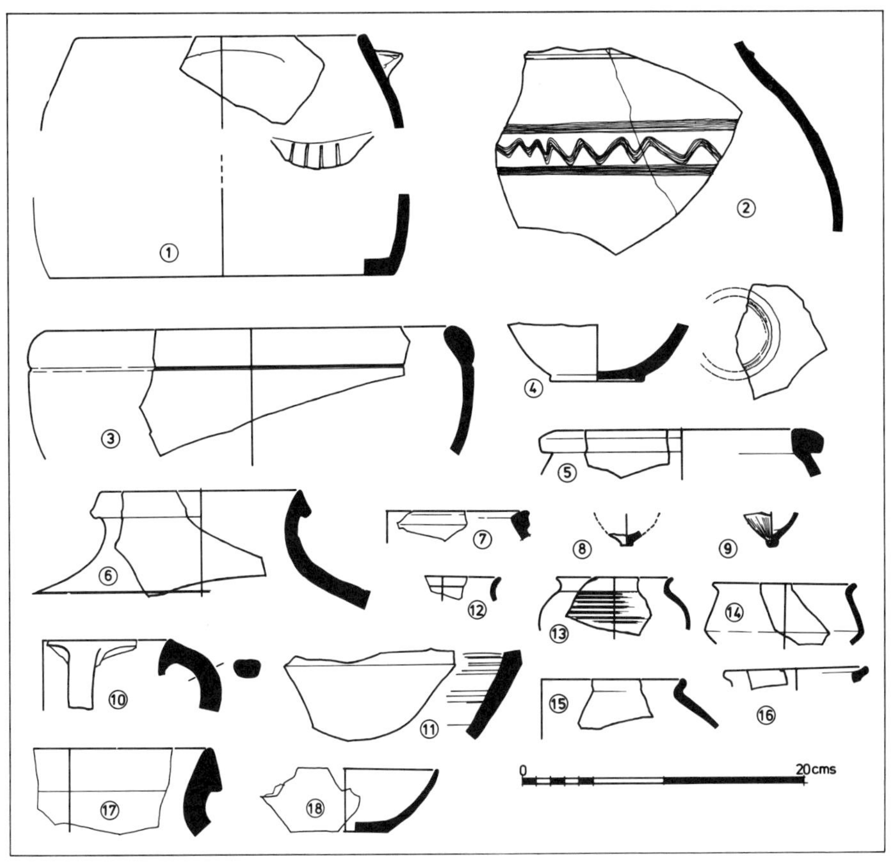

Fig. B1 Pottery: MBA Citadel complex

serving food. Types B4.4 and .5 were obviously used as lamps because oily ash marks can still be seen at their rims. Typical is type B4.5 which has a special depression to accommodate the wick. Common design features are vertically-pierced lug handles (B4.7), high loop handles (B4.1 and .2) and pushed-up lugs (B4.6). Two examples of stamped seal impressions are of special interest (B4.6). Very similar designs are reported by James Mellaart (unpublished) who says they were found at fourth millennium sites in the Jordan Valley.

Many of the pots, especially the heavier storage jars, had been mended very carefully suggesting that they did not contain liquids

Table B1 Ceramic parallels

figure		*Phase 1*	*Phase 2*	*Phase 3*	*L*	*P*	*E*	*M*	*H*	*A*
B2	1		– – – –	———	x					
	2		– – – –	———	x					
	3		———	– – – –				x		?
	4		– – – –	———					x	x
	5	– – – –	– – – –	———				x		
B4	1		– – – –	———		x				
	2	– – – –	———	– – – –		x	x	x		
	3		– – – –	———	x					
	4			———		x				
	5			———		x	x			
	6		———	– – – –				x		
	7	– – – –	– – – –	———	x					
	8		– – – –	———					?	
	9		– – – –	———			x	?	?	
	10	———	– – – –		x		x			
	11	———	– – – –						x	
B5	1		———			x	x			
	2		———			x	x			
	3	———			?					
	4	– – – –	———	– – – –	x	x	x			
	5			———	x	?	?			
	6		– – – –	———	x					
	7	———			?					
	8			———			x			
	9	———			?					
	10		———		?					
	11		———			x	x			
	12		———		?					
	13			———			x			
	14	———			x					

L Late Chalcolithic (Ghassul) (Mallon and Koeppel 1934, 1940; Perrot 1955; Hennessy 1969)
P Proto-Urban A and B (Kenyon 1979, 1960–5)
E EB1 (Amiran 1969)
M Mellaart (personal communication and 1962)
H Hama K (Fugman 1958)
A Amuq G (Braidwood and Braidwood 1960)

but cereals, and that pottery was a valuable commodity at Jawa which had to be preserved carefully like all things man-made in a marginal environment.

Table B1 demonstrates the distribution of pottery types throughout the three major phases at Jawa. The apparent absence of many types from phase 1 is probably the result of limited sampling. In general virtually all types occur to some extent in all three phases, underlining the shortness of Jawa's urbanism.

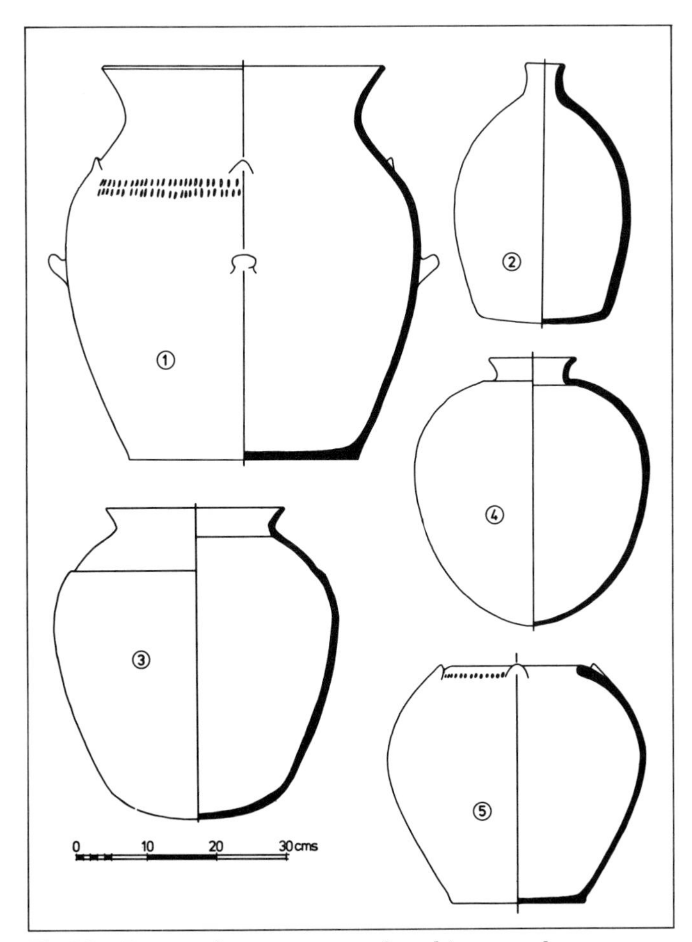

Fig. B2 Pottery: large storage and cooking vessels

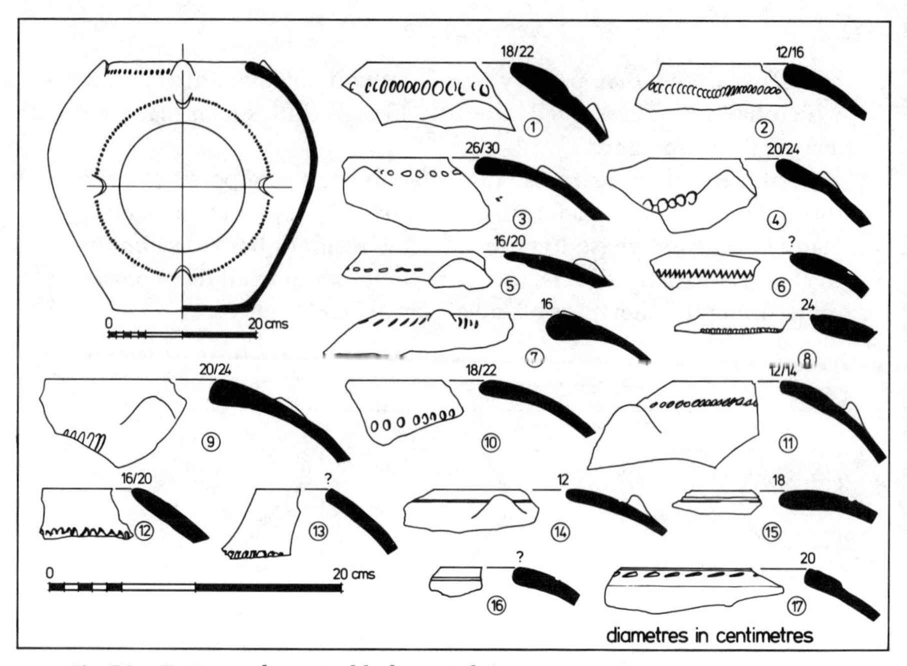

Fig. B3 Pottery : decorated hole-mouth jars

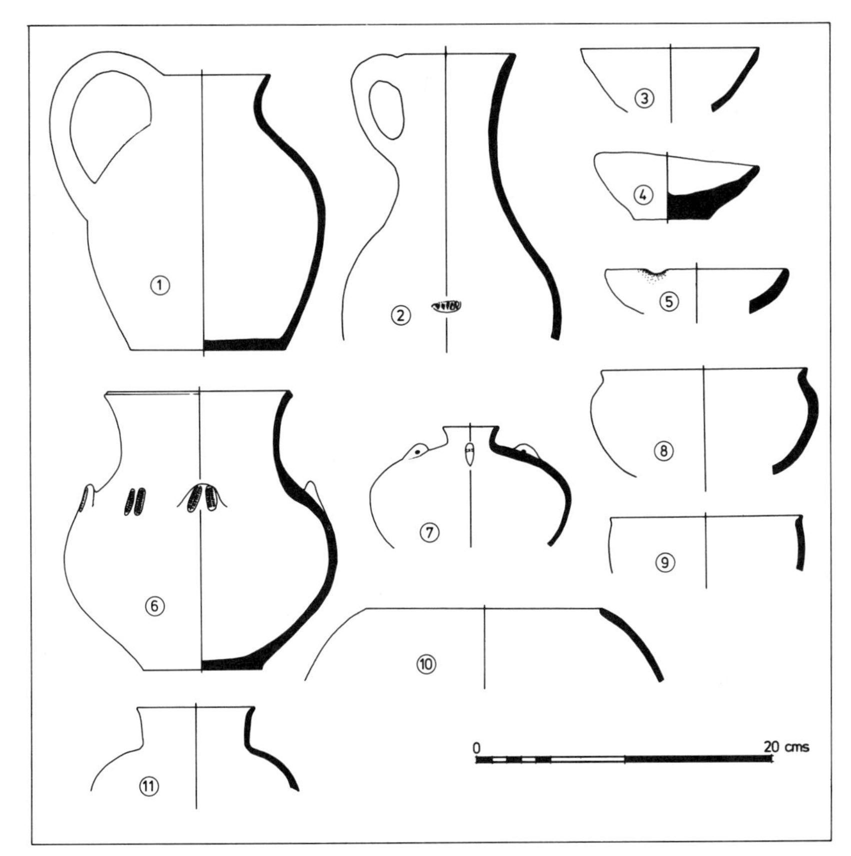

Fig. B4 Pottery: jars and cups

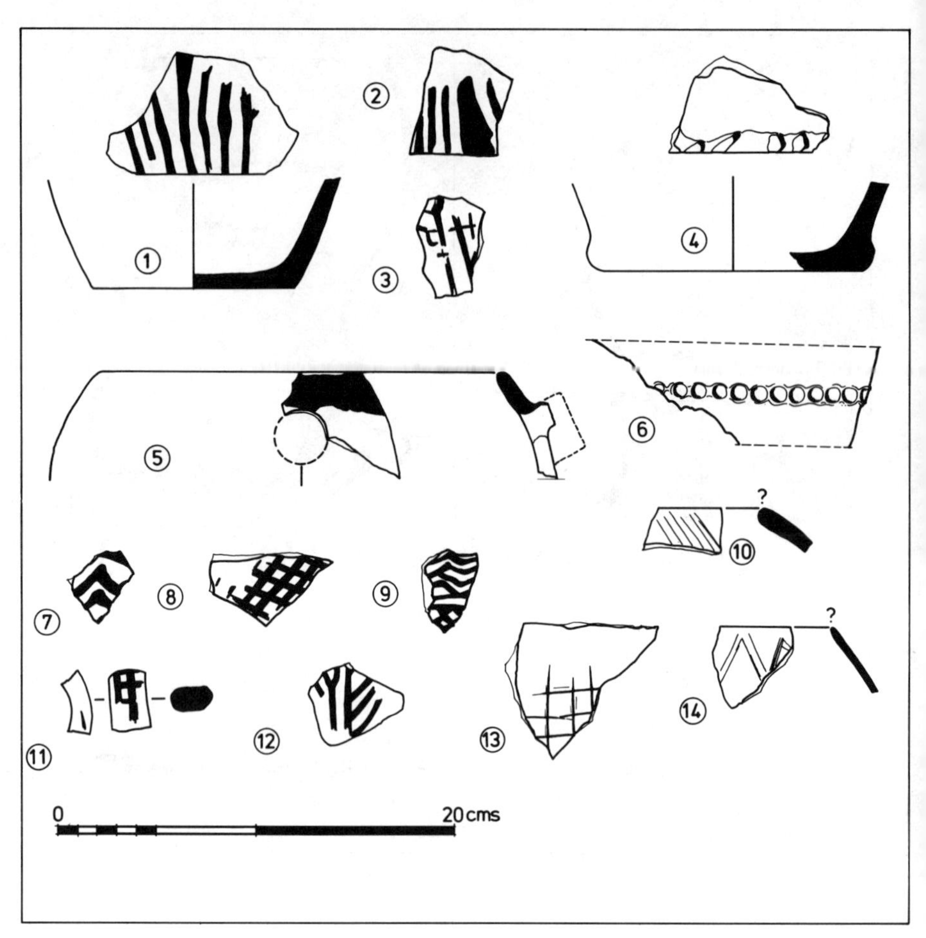

Fig. B5 Pottery: decorated pottery

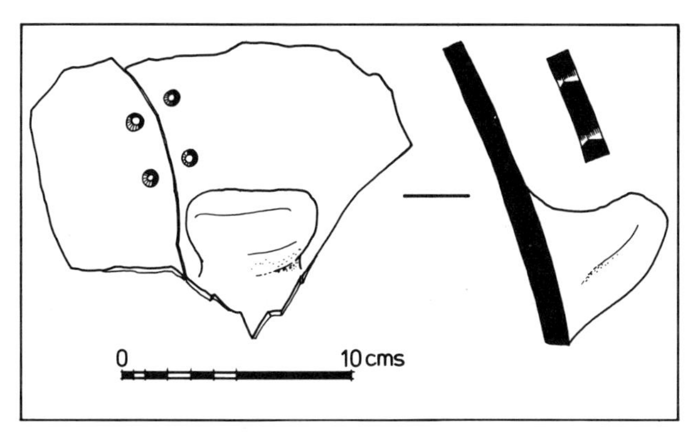

Fig. B6 Pottery: mend-holes in storage jars

2) *Flint industry*

This section is based on a report prepared by R. Duckworth. See also Duckworth in Helms (1976) and Appendix C: Qa'a Mejalla Survey (1979).

All flint found at Jawa had to be imported from outside the Black Desert, from the bast flint-strewn lands to the east (Hamada) and the south-west (Ardh es-Sawwan). Many of the flint tools had more than one use and this partly reflects the truly marginal nature of the environment in which even today the bedouin adapt tools far beyond the intentions of manufacturers. Jawa's tools accordingly served the farmer and the hunter as well as the carpenter, tanner and even the soldier. The following is a summary of the late fourth millennium tool repertoire.

Blades

Long blades, triangular or trapezoidal in section, are retouched on one side on both dorsal and ventral surfaces (figure B7.1). They bear silica sheen running the whole length of the blade at a constant width parallel to the cutting edge. Denticulation is not very coarse. These blades would have been used as knives for cutting vegetable material such as wood or reeds.

Sickle blades occur in two variants: short blades with steep retouch and non-backed blades of varied length and size (figure B7.2 and .3). Both have either triangular or trapezoidal sections and some

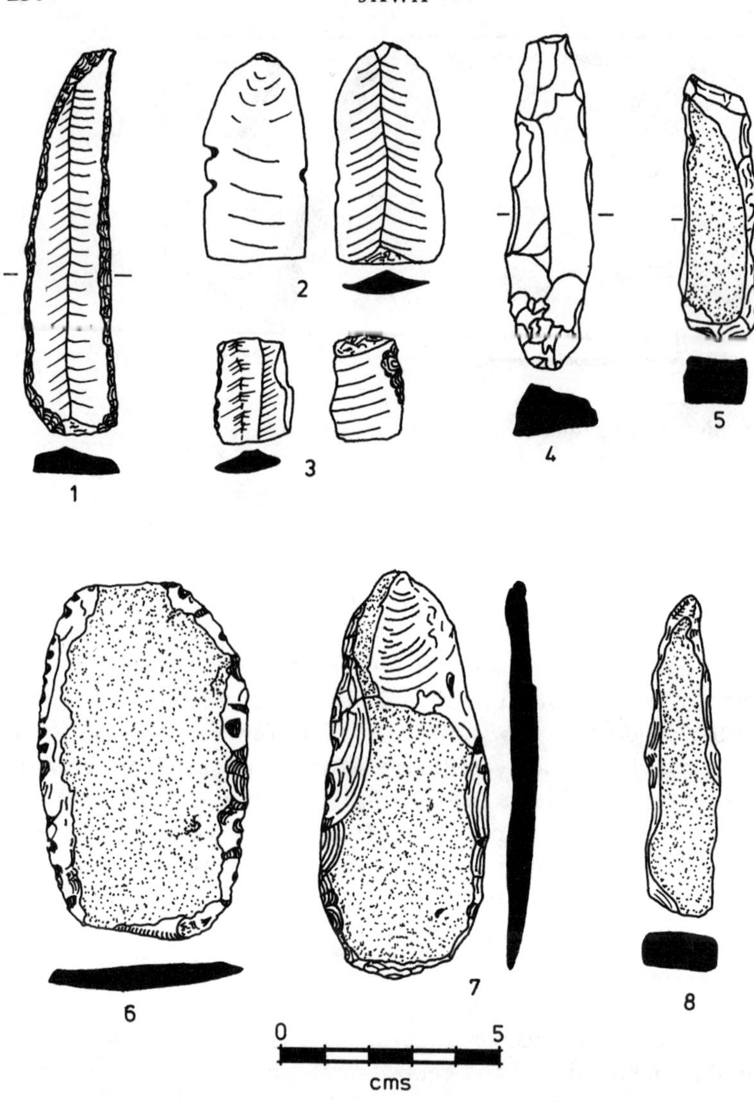

Fig. B7 Flint: 1–8

are tapered and might have been end-blades in a sickle. Denticulation is fine with retouch on both dorsal and vental surfaces. Where coarse denticulation occurs it is usually the result of secondary retouch. Silica sheen occurs, sometimes on both edges of the non-backed blades. Occasionally secondary retouch, altering the shape of the used edge, is found. In some examples the cortex of the tabular flint has been left on. Sometimes the bulbar end of a broken blade is used. Such blades tend to be notched.

Chisels and gouges

Chisels (B7.5) and gouges (B7.4) may have been used to shape wood. The former are made of tabular flint with the cortex left on the dorsal surface. The parallel sides are steeply retouched and the ends worked in an apparently irregular fashion that could be the result of use. The gouges tend to be hollow under the working edge. Their sides are roughly parallel, tending to taper towards the heel. Sections are trapezoidal and the ventral surfaces are flat. Some examples have a slight S shape resulting from secondary retouch on the sides.

Points

This is the largest group of tools found at Jawa and can be further classified according to function.

One type seems to employ a deliberate technique whereby the point is fashioned about the bulb of percussion. The natural curve of the bulb is used for one side and the other side is formed by retouching in a concave fashion. Trimming on the opposite surface to the bulb is a way of making these implements thinner. Some may have been used as projectile points; others may have been awls or borers. They were made from small round flint pebbles by chipping off a segment and then retouching the chip, using one or two blows to produce a doubly concave point.

The awls illustrate several interesting parallels in terms of shape with implements found at Shaar ha Golan in Israel dating from the Late Neolithic period. Some are made of tabular flint (B8.1–.6) and also by reshaping broken tabular scrapers (see below) by secondary retouch. They could have been used to pierce holes in leather as well as soft wood. They could even have been hafted as drill bits (B8.3) to make the mend holes in pottery storage vessels.

There are, however, definitely specialized drills or borers in the Jawa repertoire. These are made of tabular flint (B7.8). In general they are narrower and tapering with the sides steeply retouched. Some show wear that could be the result of a circular drilling motion.

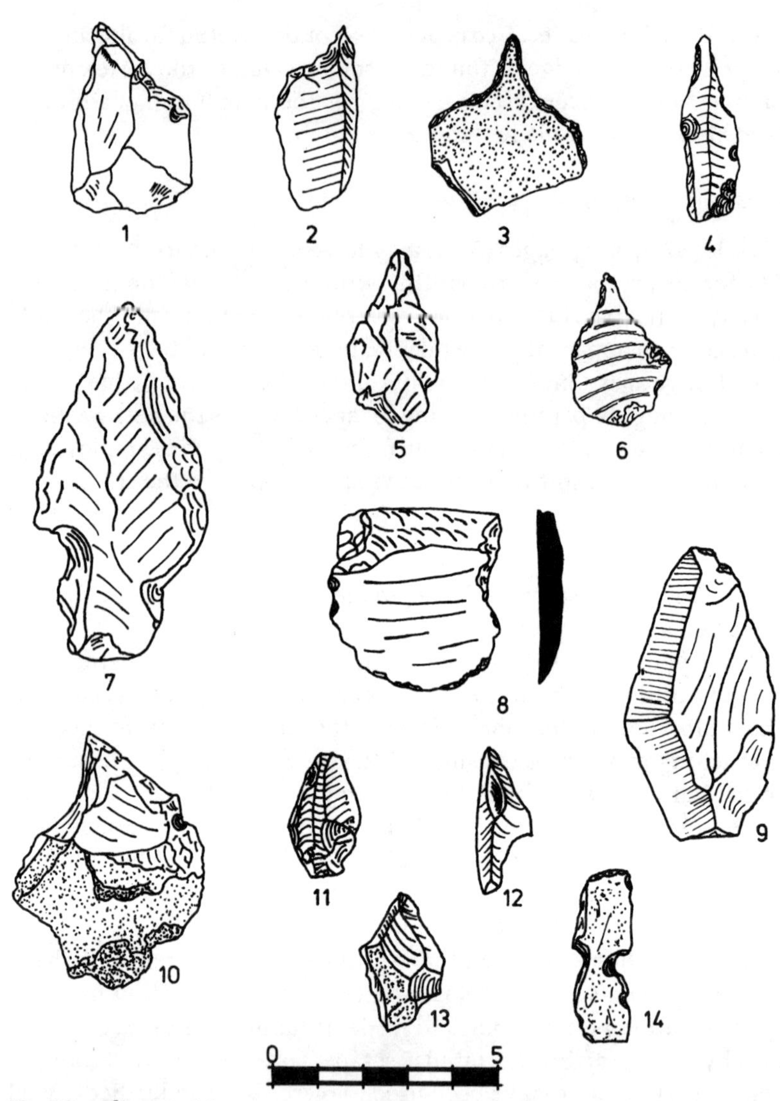

Fig. B8 Flint: 1–14

Some borers are produced by retouching and reshaping old blades. The retouching is done on both long sides to produce the elongated point.

The largest group of points at Jawa consists of projectiles (B8.7–.14). They vary in size from about 3 cm to 8 cm. In almost all cases there is no pronounced tang and only one specimen is notched on both sides. Some shapes may have parallels in the Late Neolithic period, although they are never as well made, nor do they have the typically long tangs. One large point resembles a type from Munhata in Israel (aceramic Neolithic) (B8.9) and another, longer still, may have been the head of a hunting spear. Several examples of notched or grooved, carefully-worked basalt implements have been found at Jawa (B15.1) that have been identified as arrow straighteners. They are normally associated with the aceramic Neolithic period.

Scrapers

One variety may be classified as 'fan scrapers' of a type well attested at Chalcolithic sites such as Tuleilat Ghassul. Others are elliptical, oval, rectangular and even square. They are made of tabular flint. Sometimes the bulb of percussion is trimmed down to flatten the ventral surface. On other examples the dorsal surface is trimmed. All

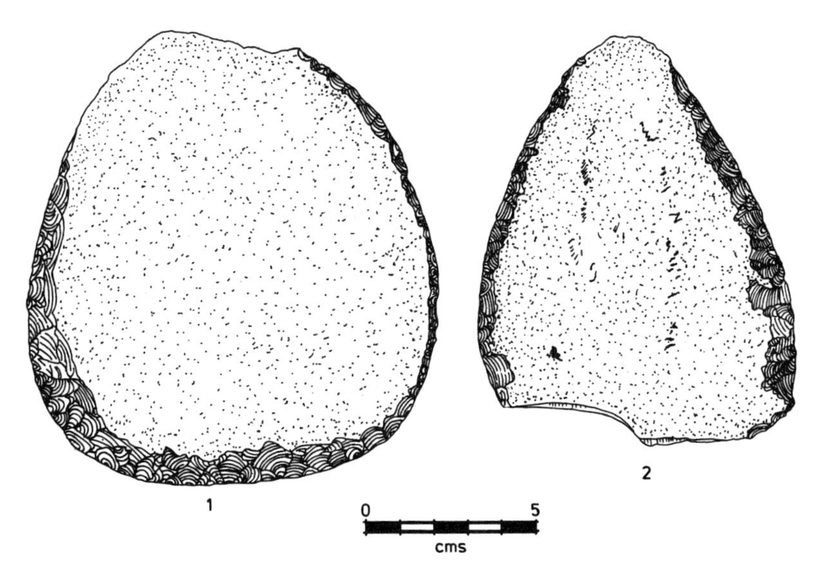

Fig. B9 Flint: 1–2

scrapers are retouched on the dorsal surface only, usually the whole way round, giving a complete working edge apart from the bulbar end (B7.6 and .7 and B9.1 and .2). Some scrapers have silica sheen.

The function of scrapers has occasionally eluded researchers, but it seems quite reasonable that besides being used in scraping hides they could also have served as adzes, planes, bark strippers and even hoes.

The tool kit assembled here developed out of a long line going back to the Neolithic period. The tools have become simpler and less specialized. There appears to be an increase in multi-purpose implements, but still a continuation of function and the activities they imply. Apart however from offering a view of Jawa's handicrafts this assemblage also serves to confirm the date derived from the pottery in Appendix B. Parallels may be cited for most of what has been found and they all point to the latter half of the fourth millennium BC: definitely prior to the Early Bronze Age of Palestine.

To summarize, with the exception of certain blades (very few), sickle blades similar to those of Jawa have been found in the deep sounding at Nineveh in northern Mesopotamia – where the 'roots' of the Jawaites may be sought – and many of the points have been shown to derive from Late Neolithic types. Scrapers resemble material from Chalcolithic sites in the Jordan Valley. In general the flint implements described here can be dated in the latter part of the fourth millennium, towards the very end of the Chalcolithic period. Only some blades, the first noted above, might refer to the so-called Canaanean culture that is more commonly associated with the Early Bronze Age of Palestine to the west. The rest also appear at Ghassul (period IV). They are also found at 'Affuleh in the Easdraelon Plain south of Galilee, the wadi Ghazzeh sites, at Horvat Beter, Jericho, Beisan, Umm Qatafa and Hama in Syria. All of these parallels point to the later fourth millennium BC. For reference see Koeppel (1940), Neuville (1940), Crowfoot (1937, 1948), Neuville and Mallon (1931) and FitzGerald (1934).

3) *Ground stone objects and others*

Jawa's repertoire of ground stone objects is not large, but it is comprehensive and illustrates most of the domestic activities that took place in and about the ancient town. However, for dating purposes stone objects are very limited. The style of some originates in much earlier periods and continues, often without much change,

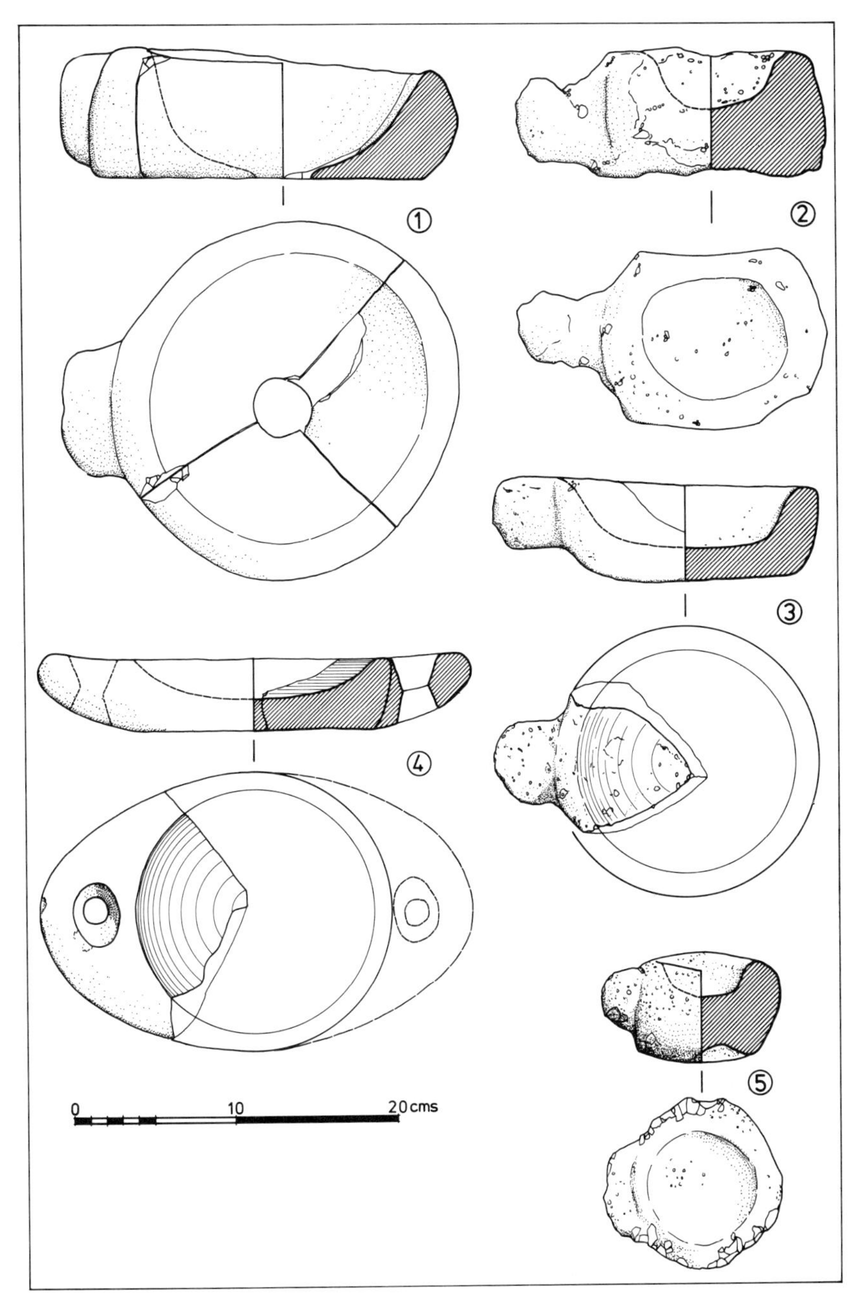

Fig. B10 Ground stone: vessels with handles

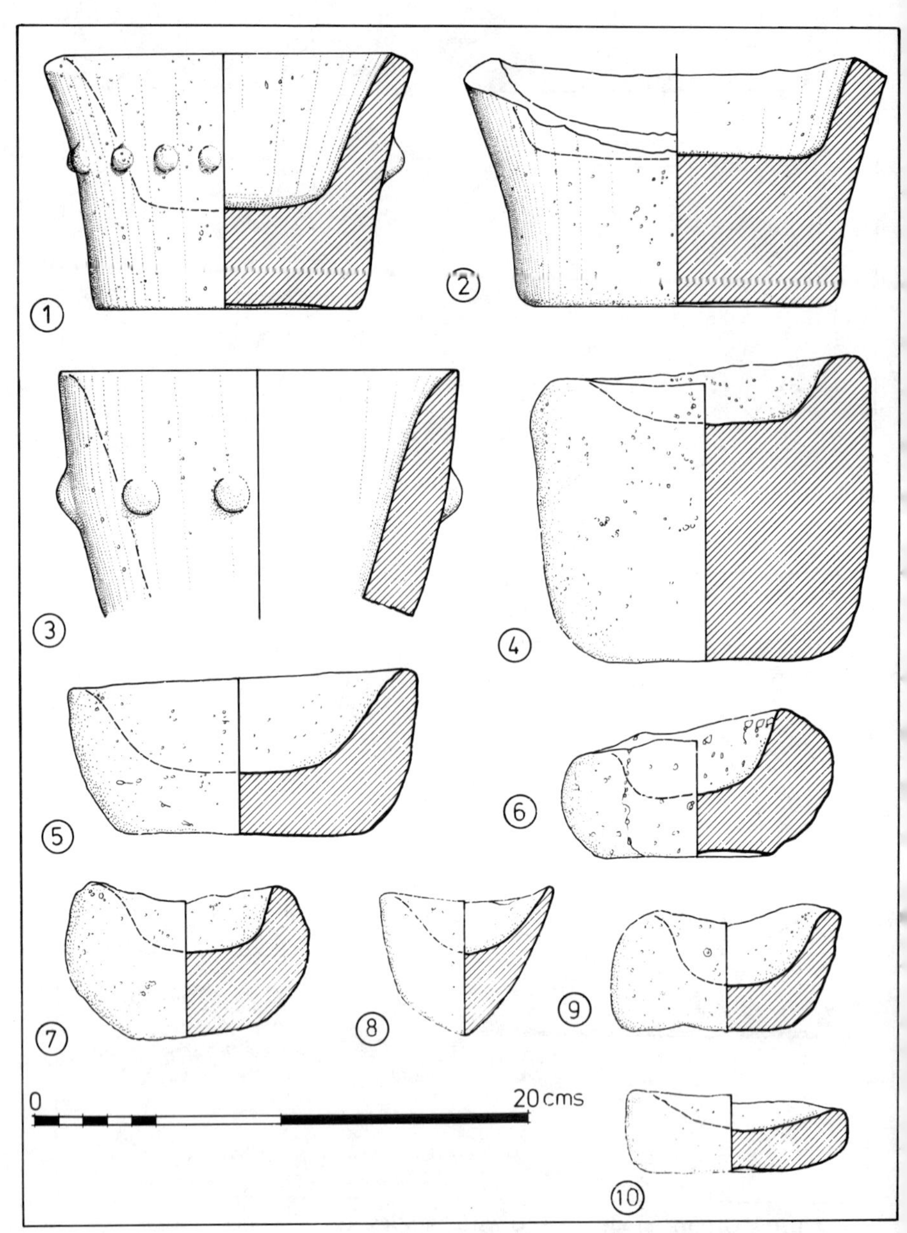

Fig. B11 Ground stone: bowls and mortars

well beyond the time span set for urban Jawa. Only two groups can be dated more closely and these are the ground stone bowls decorated with bulbous protrusions (B11·1–·3) and – with some qualifications – the polished mace-heads (B14·1–·4). The bowls have been found at various sites in Palestine and date from the so-called Proto-Urban period or the early part of EB1. They also have some affinity with certain basalt containers found at Chalcolithic sites throughout the Levant. A relationship may exist between the decoration of those bowls and a type of grey burnished pottery found in Palestine (Esdraelon ware) likewise attributed to the beginning of EB1 (or Proto-Urban C). The mace-heads, on the other hand, are widely known throughout the Near and Middle East and so far as tracing stylistic connection to and from Jawa is concerned their distribution is of little value. Furthermore their use is attested well into the third millennium, thus also limiting their dating value. Maces are known in Egypt from predynastic times (fourth millennium) onwards and also appear in contemporary sculpture as the 'royal' symbols of power and warfare. The most famous of these is the palette of King Narmer and an actual mace-head depicting another predynastic king is illustrated at the beginning of Section IV. Maces of this general type are also found in Syria (Hama) and farther north in the Amuq Plain (period G: late fourth millennium). They have also appeared in similarly-dated assemblages in Mesopotamia, in Anatolia and even in the virtually undated rock art at Jawa and in Arabia. In terms of providing a date for the Jawa maces these parallels would only support a *terminus post quem* in the late fourth millennium BC, a date in keeping with the rest of the artefactual evidence.

Nearly all of the stone objects were made by abrasion of stone upon stone, either by hand or crude 'machine'. This means that a relatively advanced form of the drill was used at Jawa – perhaps one working on the torsion principle. However, the possibility of importation of finished objects cannot be ruled out; particularly in view of the migration hypothesis set out earlier in the book, and the obvious brevity of Jawa's urbanism.

The functions of the ground stone objects, apart from the manufacture of tools, are easily recognized and complement other aspects of Jawa's daily life. The vast majority of implements were used for food processing. Saddle querns (B13·2) are the most common

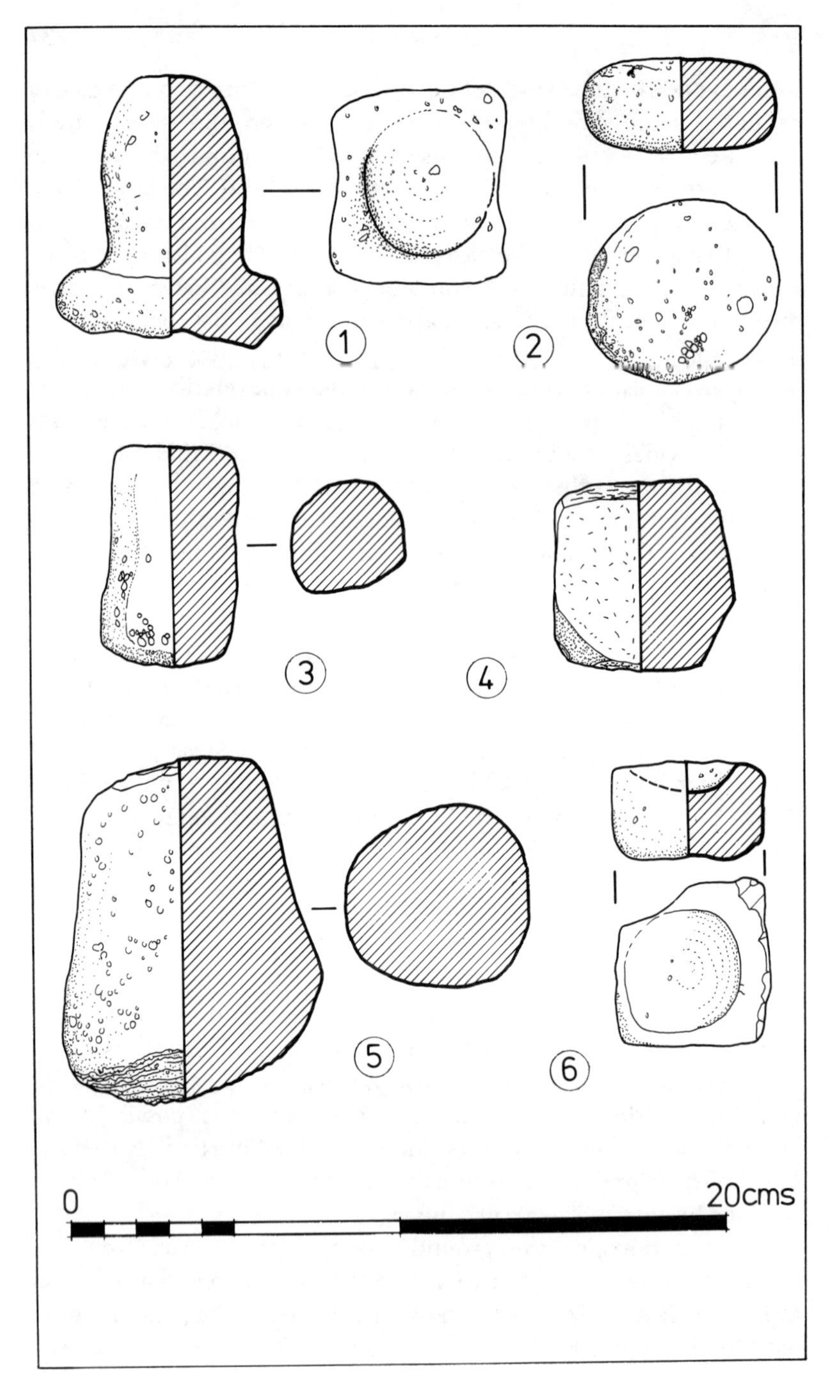

Fig. B12 Ground stone: pounders

and would have been used, as they are today, to grind seeds with a hand-held stone (B12·2). Two sizes occur: a 'household' quern 20–40 cm in length and a less common 'communal' quern of 60–80 cm. A more refined grinder bowl is represented in figure B10·1. It has a hole in the bottom which was probably caused by wear rather than being intended for any purpose. Grinding would have been done with either a stone pestle (B12·5) or a larger, rounded stone with an offset handle.

The logical conclusion derived from these objects is that food processing was the major domestic activity of the people at Jawa and that much of what they ate had to be ground down, pounded and pulverized. All of this fits well into the subsistence pattern normal to the region, as it is today and as it was when the town's life-support systems were functioning. The abundance of grinders, specifically the proportion of small ones to larger, also shows that food processing – as well as storage, as has been seen – was not communal. Each household had its own full complement of tools. The manufacture of food then was a domestic industry and apparently as decentralized as storage.

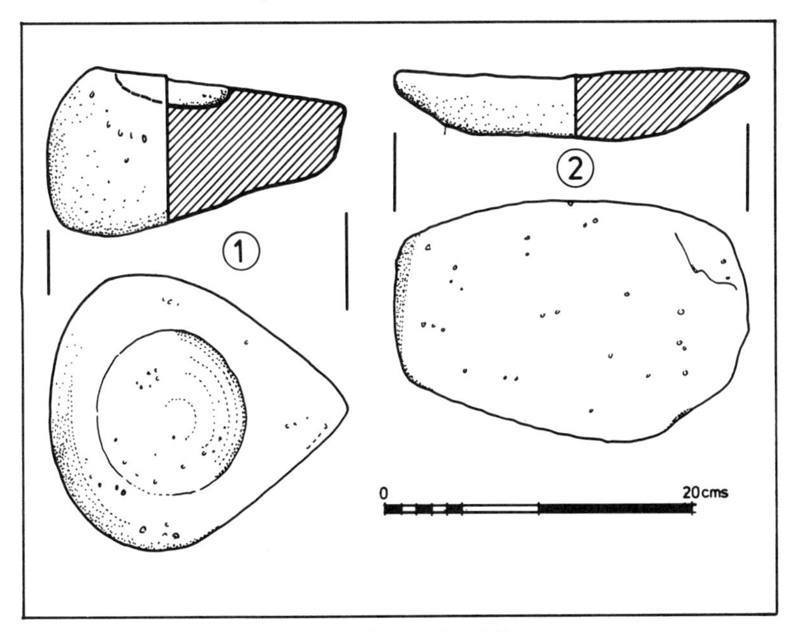

Fig. B13 Ground stone: door socket and saddle quern

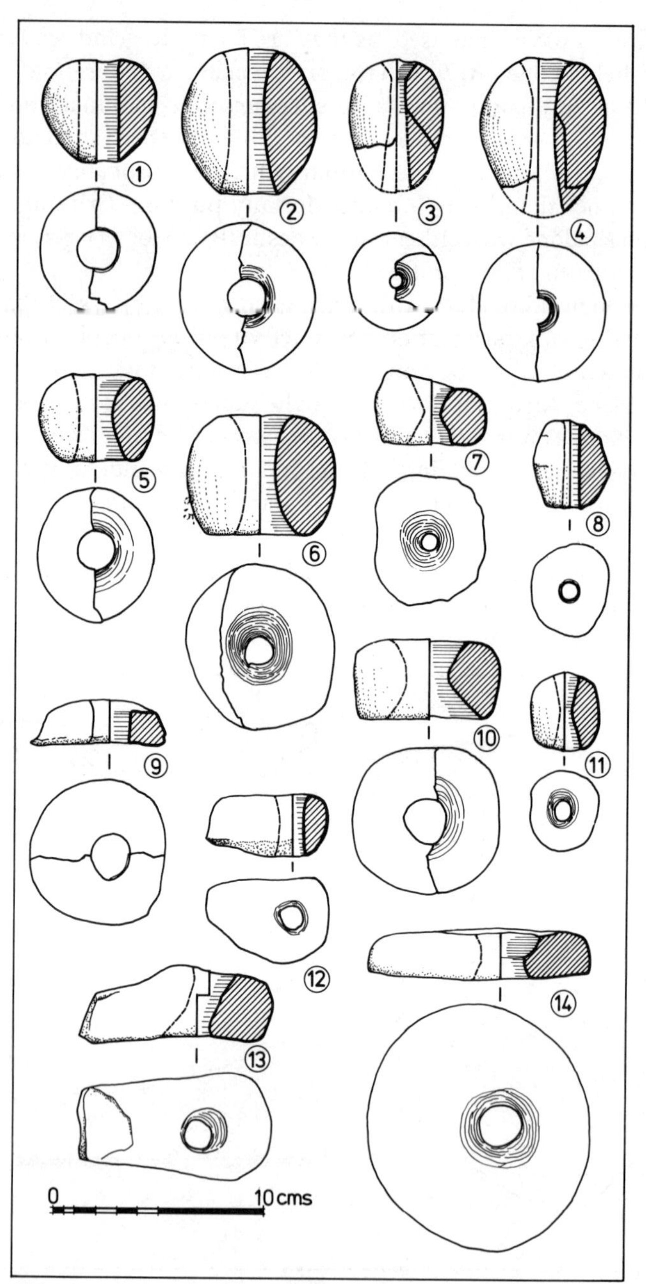

Fig. B14 Ground stone: maces and hoes

Whereas the objects discussed so far represent home economy the balance could indicate specialized industry on a modest scale. This must be qualified. On the whole we did not find workshops at the town site and one reason for this – apart from limited excavation – may be that they lay outside the town as is normal for example with flint-knapping sites. The same might apply to brickworks, wood cutting and other industries whose raw materials were bulky, were imported from abroad and might therefore have been specialized activities under general community supervision. On the other hand workshops are not so easily recognized on architectural grounds alone unless a good assortment of special tools and installations as well as some raw materials have survived. Jawa was by no means a rich town according to what has been found.

This rather negative evidence is bolstered by the discovery of the two 'kilns' in area F and on this basis it may be possible to speculate that the objects under discussion were made in specialized workshops at Jawa. This is a view that is not entirely discordant with the generally high level of technology at the site.

Objects like the door sockets (B13·1) are very common and always of the same dimensions and design. They would have been made by 'professional' stone grinders in the same way as wood might have been cut into suitable lengths for the domestic structures. The

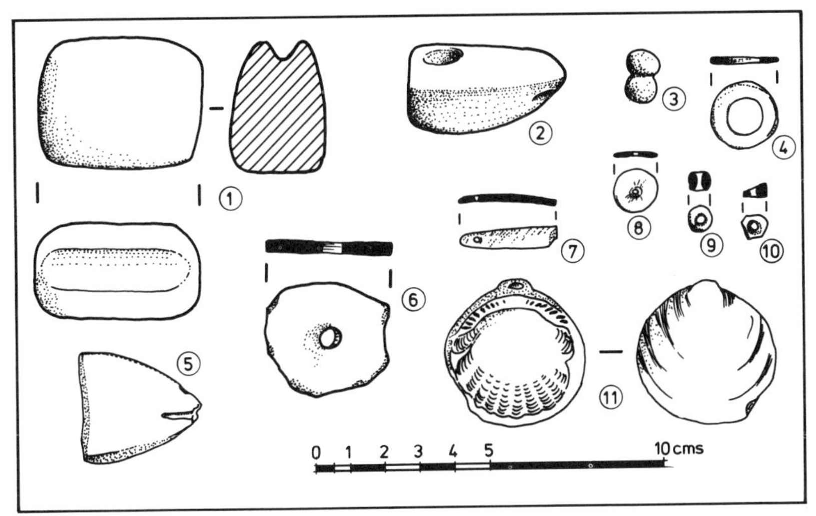

Fig. B15 Small carved and drilled objects

highly-polished finish of the maces betrays a high level of artisanship, as does the treatment of some of the stone bowls (B10·4). Just as the presence at Jawa of trained engineers is evidenced by the excellence of the water systems so the frail evidence for craft specialization might point to professional classes at ancient Jawa.

Drilled and polished stone objects are mostly of a domestic and/or small-industry nature: either for the manufacture of tools or, like figure B14·8 and ·11, for use in the home as spindle whorls and loom weights. Only two groups of objects go beyond that: one as a natural corollary to food production, the other to warfare. The first is represented by figures B14·12 and ·13 (the latter half-scale) and can be identified as hoes. Alternatively these tools may have been used as crude adzes or even weapons.

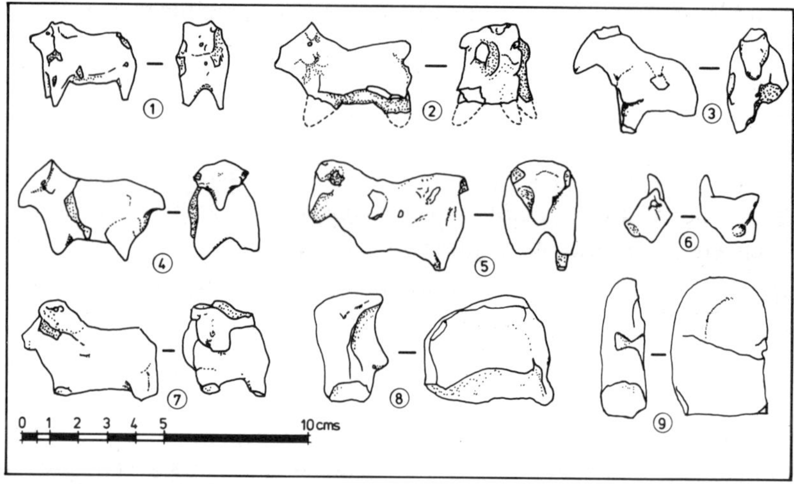

Fig. B16 Unbaked clay animal figurines

Figures B15 and B16 illustrate miscellaneous items that show how little was left in the houses of Jawa when the town ceased to function. They are all that we found. A few pierced beads, several (broken) decorated bone pins, assorted marine shells and crude animal figurines (sheep/goat?) are all that relieve the purely functional aspect of ancient Jawa.

APPENDIX C

QA'A MEJAḶLA SURVEY (1979)

by A. V. G. Betts

Two surveys were carried out in the basalt region south of Jawa to study the prehistoric occupation of the desert. Both survey areas lie on wadi Rajil: one some 50 km east of the oasis el-Azraq and the second further to the north (figure 17). Prehistoric sites are numerous in both areas, occurring mainly on the edge of the basalt where it projects into sand and mudflats or, in the case of the southern area, into the surrounding flint desert. The sites can be roughly divided into seven categories. Symbols in brackets refer to these sites as described in the main text (chapter 6).

1) Hut-circles/corrals without flint (C2)

These are irregular clusters of sub-circular structures. They vary in size, complexity and also location, some lying directly on the basalt/sandflat margin, some on sheltered slopes and others on the tops of promontories. It is likely that they vary considerably in date and some have been re-used by the modern bedouin.

2) Hut-circles/corrals with flint (C2)

These show the same variety as (1) above. Much of the flint is undiagnostic, but recognizable Neolithic artefacts are associated with some of the sites.

3) Modern bedouin sites (B)

These divide into two types: camp sites with hearths and lines of stones to hold down tents and hut-circles/corrals, some clearly recent constructions and others re-using earlier structures.

4)'Jellyfish' (C1)

These regularly-shaped sites 20–50 m across usually consist of two concentric circles of low stone walling divided into irregular segments by radiating walls. Some have hut-circles attached to the outer ring.

5) Isolated huts/cairns

Cairns are usually associated with Safaitic inscriptions, but there are some which could possibly be linked with the Neolithic period. There are also clusters of individual hut-circles, often no more than 3–4 m across and sometimes hard to distinguish from cairns.

6) The 'Desert Kites'

See chapter 6 for description.

7) Flint knapping sites

These are found mostly on hill tops and include scatters of waste, blanks, rough-outs and croes. The majority of these sites can be assigned to the Neolithic period. They are not associated with architecture of any kind.

Flint artefacts

On most sites flint artefacts are rare or non-existent. The reason for this becomes clear when it is remembered that these sites lie in a marginal area and were used by nomadic populations who rarely leave much behind. However, three types of site have yielded enough to provide dating evidence. Category (7) sites, the flint-knapping floors, are mainly for the production of tangled, sometimes pressure-flaked points on blades struck from bi-polar cores of sub 'naviform' type. Similar pieces found in the Azraq region have been ascribed to the Pre-pottery Neolithic or possibly Late Neolithic period (Copeland, in Garrard *et al.* 1977).

Most of the sites in category (2) are undatable but there are a few with thick surfaces scatters of artefacts. An extraordinarily high

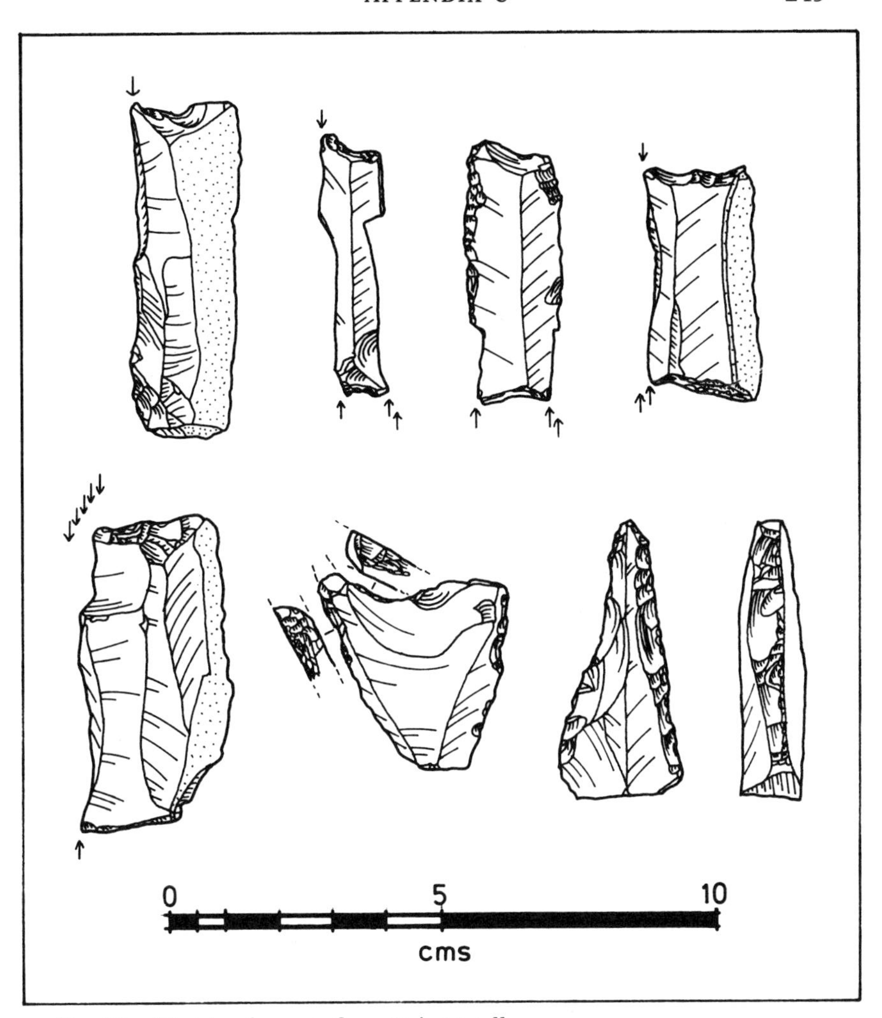

Fig. C1 Flint implements from Qa'a Mejalla survey

proportion of the tools on these sites consists of concave truncated burins (figure C1). Other tools include tabular scrapers, awls and drills. The density of artefacts on these sites, especially in comparison with the barrenness of the majority, seems to indicate some form of semi-permanent occupation. The date of these sites is as yet uncertain due to a lack of excavated material. Similar truncation burins have been found at other desert sites, notably at wadi Dhobai

(Waechter and Seton-Williams 1938) and at Jebel Umm Wu'āl (Field 1960). These have been ascribed to the Pre-Pottery Neolithic B period. However, the tabular scrapers are of a type known in the Late Neolithic (Moore 1973) and indeed one example is close to material from the Jawa Late Chalcolithic assemblage (figure B7).

The third group of sites with diagnostic artefacts is category (6), the 'Desert Kites'. One 'kite' in the southern survey area yielded a cache of four 'lance points' of a distinctive rose-pink flint with quite fine bi-facial retouch (figure 23). These bear a resemblance to Neolithic Byblos points and were found lying in a 'hide' at the head of the 'kite'. Several of the 'kites' in the northern survey area produced examples of tanged and fluted-retouch arrowheads rather similar to Late Neolithic Amuq points, but also having parallels in earlier Pre-Pottery Neolithic examples. Similar arrowheads have been found at Azraq.

One of the most surprising facts revealed by these two surveys is the abundance of prehistoric sites in a marginal area such as the Black Desert. Of these perhaps the most fantastic are the 'kites' as their size and number are clear indications of a high level of sophistication in social organization and hunting techniques among ostensibly 'Neolithic bedouin'.

APPENDIX D

PLANT REMAINS

by G. H. Willcox

Samples were collected from areas of blackened earth in trench II of area LF belonging to phase 2. The samples were prepared for macrobotanical analysis by simple dry-sieving using a 1 mm aperture mesh. These preliminary investigations revealed the presence of seeds of the following taxa:

Hordeum vulgare L. emend. LAM	six row hulled barley
Triticum monococcum L.	einkorn
Triticum aestivum sensu lato	bread wheat
Triticum dicoccum (SCHRANK) SCHÜBL.	emmer
Cicer arietinum L.	chickpea
Vicia ervilia (L.) WILLD.	bitter vetch
Lens culinare MEDICUS	lentil
Pisum sativum L.	pea
Galium spp.	bedstraw
Vitis vinifera L.	grape
Crataegus cf. monogyna	hawthorn

In addition pieces of charcoal of *Rhamnus* (buckthorn), two small fragments of *Quercus* (oak), and a bushy chenopod (probably *Kochia*) were identified.

The wheat grains were mainly runted specimens and caused difficulties in identification. The bread wheat grains resembled those of *Triticum compactum* HOST., but in the absence of rhachis material a definite identification is not justifiable. Seeds of *Galium* and *Triticum monococcum* were rare, and einkorn may have occurred as a weed rather than as a crop in its own right.

This group of crop plants could only have been grown at Jawa under present conditions in a year of exceptionally favourable rainfall: irrigation agriculture seems to be implied, as might be expected. The presence of oak charcoal is interesting in view of the later biblical references to this wood in the Jebel Druze (biblical Bashan).

APPENDIX E

ANIMAL REMAINS

by Ilse Köhler

The following comments are the result of a preliminary study aimed at determining the major animals at ancient Jawa. While the list of species is not yet complete (a few non-domestic small animal bones have not yet been identified) the percentages given here will not be altered significantly.

Table E1 Faunal remains: actual finds, relative importance

		No. bone fragments	%
Ovis aries	domestic sheep	2206	86·7
Capra hircus	domestic goat		
Gazella spp.	gazelles	60	2·3
Bos taurus	domestic cattle	217	8·5
Equus spp.	equids	53	2·1
Canis sp.	dog	1	
Lepus sp.	hare	4	0·4
Gallus sp.	fowl	3	

Sheep/goat

The preponderance of sheep and goat bones is not unexpected since this is usually the case at Near Eastern sites. From their earliest domestication up to the present day these small ruminants have been the most frequently kept animals. The varied uses to which they can be put (production of meat, milk, wool and hides) as well as their hardiness and the advantages of their size accounts for this.

While sheep or goats are basically as efficient as cattle their smaller size gives them the advantage in regions of poor grazing since they have to forage over smaller areas of ground per day. In poor communities they also have the advantage that less meat is produced at each individual slaughtering – an important factor when no means of preservation was available.

The sheep to goat ratio is always very important since these two species differ in their feeding habits. While sheep need to graze the goats browse on a wider variety of plants including bushes and tree-tips, so that a shift in their relative proportion is often interpreted as an indicator of environmental change. Under good conditions sheep predominate (their carcass is preferable to that of the goat) whereas in dryer times the relative number of goats increases.

The accurate differentiation of goat and sheep bones is problematic. This is due to the great morphological similarities in most parts of their skeletons which is why they have been enumerated and discussed together. In this study the sheep to goat ratio was estimated to be 4:1 by using the distal ends of the metapodial bones as the criterion.

Cattle

While cattle bones represent only about 10 per cent of those from sheep and goat it must be remembered that cattle produce ten to thirteen times as much per head and so it is obvious that beef played an important role in the diet of the Jawaites. Thus milk, hides and possibly also physical power may have contributed substantially to the economy.

Jawa's cattle had long horns and were of considerable size; almost within the size range of wild cattle (aurochs). Further study may show aurochs were actually part of the wild fauna at Jawa. However, if that were the case it would indicate a major climate change which seems improbable.

Jawa's zoomorphic rock drawings (see chapter 5) can now be linked tentatively with the fourth millennium town. Furthermore the drawings may depict domesticated cattle, if one regards the characteristic segmentation of their bodies as 'piebald' markings, typical of domesticated animals. Such markings do not occur in wild populations.

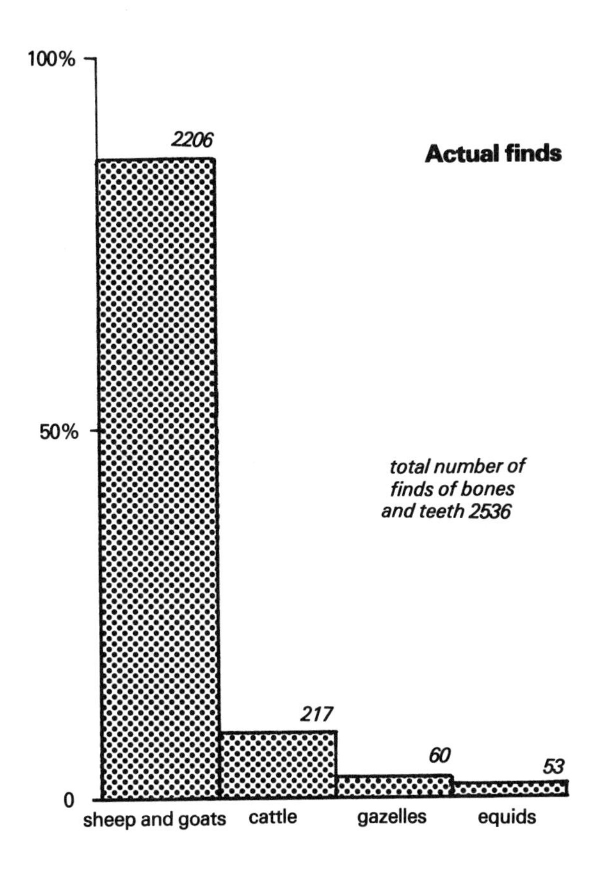

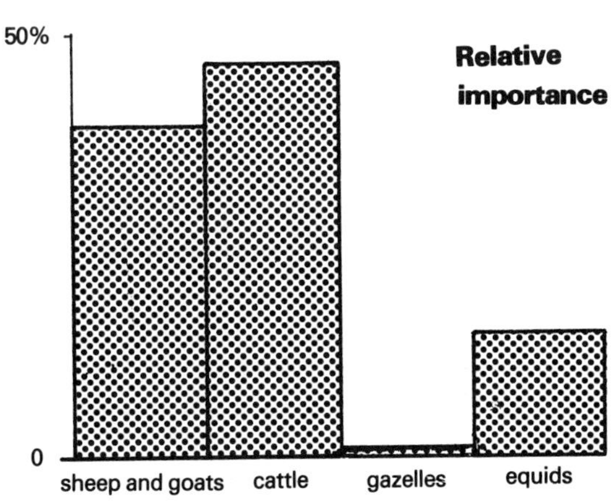

Fig. E1 Graphs: actual finds and relative importance

Gazelle

Only sixty bones were attributed to gazelle and although allowances must always be made for a few fragments being undifferentiable from those of other small ruminants it can be established that they were not a dietarily important animal and that gazelle hunting was not of great economic significance. However, remains of 'kites' used to trap animals can still be seen in the immediate vicinity of Jawa. Gazelle hunting might have been more a prestige-earning activity and leisure occuation than a necessity.

Equids

At the present stage of the investigation the equid remains from Jawa can only be tentatively identified as the Syrian Onager (*Equus hemionnus hemippus*) and/or the true species of ass (*Equus asinus*).

Some of the bones show chopping marks indicating that these equids were eaten. This, however, does not answer the question of whether they come from hunted or husbanded animals. The generally accepted date for the first appearance of the domesticated donkey in the Near East is in the Early Bronze Age (*c.* 3000–2300 BC), but the recent discovery of basket-bearing donkey figurines at sites roughly contemporary with Jawa (Kaplan 1969) suggests an earlier date. One of the major aims of further research will be to establish whether the equid bones are from wild ass and *Equus hemionnus hemippus* or from domestic donkeys.

Conclusion

1) The people of Jawa were farmers and kept domestic sheep, goats and cattle.

2) Cattle husbandry was equally as important as that of sheep/goat.

3) The diet was to a small extent supplemented by hunting of gazelles.

4) The extent of animal husbandry demanded that the water systems functioned efficiently (see Section V). The prevalence of sheep and cattle remains (relative to goat) implies irrigation to improve grazing.

5) Equids also played a role in the diet of Jawa's inhabitants, but it remains unclear whether they were hunted or domesticated.

APPENDIX F

MATERIAL REQUIREMENTS AND LABOUR-TIME

Weight extimates to the nearest 100 tons are based on the reconstructed volume of the various structures and the following densities of building materials:

basalt (b)	2·0–2·4 gm per cm^3
gravels (g)	2·1–2·5 gm per cm^3
soils (s)	1·6–1·8 gm per cm^3

Labour time estimates are calculated on the basis of the most efficient rates recorded of Near Eastern excavation workers (*c.* 0·59 tons per man per day) and then adjusted according to specific tasks, taking into account distance and height over and to which materials had to be moved as well as the unit weight of some building blocks. The following rates are used:

small-scale building (houses, canals)	0·59 tons per man per day
large-scale fortifications (UT)	0·16 tons per man per day
large-scale fortifications (LT)	0·32 tons per man per day
large-scale water works (dams)	0·25 tons per man per day

The labour-time estimates (months) for the various elements that made up urban Jawa are shown in brackets (0·0) to the left of the weights and are calculated on the basis of roughly one-third of the estimated Jawaite population (*c.* 700 workers) being available at the beginning of construction in phase 1. After phase 2(a) and *Compromise 1* additional bedouin labour was available and the work-time shown in figure 94 is adjusted accordingly.

Table F1 Building materials

1) *The town*

UPPER TOWN:						
Fortifications			12,000	(3·6)		
Houses etc.			49,000	(4·0)		
			61,000	(7·6)		61,000
LOWER TOWN:						
Fortifications						
LTS	10,100	(1·5)				
LTW	9700	(1·5)				
LTE	16,200	(2·4)				
	36,000	(5·4)	36,000			
Houses etc.						
LTS	16,800	(1·4)				
LTW	16,200	(1·3)				
LTE	27,000	(2·1)				
	60,000	(4·8)	60,000			
			96,000	(10·2)		96,000

Table F1 Building materials (cont.)

2) *Water systems*				
SYSTEM I:				
DaI (a)	55,000	(b)	(10·5)	55,000
(b)	35,000	(bg)	(6·7)	35,000
(c)	40,000	(bs)	(7·6)	40,000
Canals	4000	(b)	(0·3)	4000
P 01	500	(b)	(0·1)	500
P 02	3900	(b)	(1·5)	7700
	3800	(s)		
P 03	6600	(b)	(4·6)	24,400
	17,800	(s)		
P 04	3800	(b)	(2·7)	14,400
	10,600	(s)		
P 05	3000	(b)	(0·6)	3200
	200	(s)		

(a)	109,200 (20·3)	(a)	109,200
(b)	89,200 (16·5)	(b)	89,200
(c)	94,200 (17·4)	(c)	94,200

(continued overleaf)

Table F1 Building materials (cont.)

2) *Water systems (cont.)*						
SYSTEM II:						
DaII	600	(b)	(0·1)	600		
Canals	2600	(b)	(0·2)	2600		
P 06	2600	(b)				
	200	(s)	(0·5)	2800		
P 07	1700	(b)				
	200	(s)	(0·4)	1900		
				7900	(1·2)	7900
SYSTEM III:						
DaIII	600	(b)	(0·1)	600		
Canals	13,400	(b)	(1·1)	13,400		
P 08	1400	(b)				
	200	(s)	(0·3)	1600		
P 09	1200	(b)				
	200	(s)	(0·3)	1400		
P 10	1500	(b)				
	200	(s)	(0·3)	1700		
				18,700	(2·1)	18,700
						(a) 292,800
						(b) 272,800
						(c) 277,800

APPENDIX G

NOTES ON SOME SAFAITIC INSCRIPTIONS

by M. A. C. Macdonald

1) The 'Jawa Rider' (figure 16)

Preliminary study of the inscriptions suggests the following readings. (Of the two stones only the 'Jawa Rider' was read directly; the wadi Ghadaf stone was deciphered with reference to photographs alone.)

(i) l msl̊k̊ bn bdbl bn slm w ẖṭṭ (By MS[LK] son of BDBL son of SLM and he wrote [or 'drew'])

This text which is the longest of the three on this stone, runs along the lefthand side of the rock. With the exception of the *l* and the *k* of the first name all the letters are clear. Both the *l* and the cross-stroke of the *k* are more lightly scratched than the rest of the text and could be accidental. *MSLK* is so far unattested, though *SLK* is known. If the two doubtful scratches are ignored *MSB* is also unattested as a personal name, though it has been found as a tribal name (*CIS* 2702). *MSK* and *MSLB* are known. The other names are common, though in the past *BDBL* seems always to have been read as *BDRL*.

The end of the inscription is particularly interesting since it is the first clear example in Safaitic of ẖṭṭ used as a verb. As far as I know, there are four other texts where ẖṭṭ or ẖṭ could possibly be a verb, but in none of them is this certain. The exact meaning of ẖṭṭ in Safaitic, either as a noun or a verb, still lacks precision. The basic meaning in Classical Arabic of 'to make lines or marks' could equally apply to writing or drawing and it is possible that in Safaitic the word was used for both. Certainly it appears in texts which are accompanied by drawings (as in the present text) and those that are not.

It is interesting and quite typical that the two other texts on the stone could also be interpreted as laying claim to the drawing: (ii) an author of the same name, *BDBL*, though apparently not a relative, and (iii) *RB'L*.

(ii) l bdbl bn q'ṣn
The inscrition is lightly scratched to the left of (i). Both names are known.

(iii) l rb̊'l
The inscription which is extremely faint is written vertically to the left of the horseman's head. *RB'L* is a well-known name.

2) The wadi Ghadaf inscriptions (figure 25)

(i) l qṭ bn jm't

If the text continued beyond *jm't* the rest has been lost when the rock was broken at this point. Both names are known, though *qṭ* has so far been found only twice before and *jm't* only once.

(ii) l h̊l' bn ẖrj

Both names are known, though *hl'* has been found only once before and that in Thamudic (*TIJ* 58, where it is read *hn'*, though *hl'* seems the more probable reading). The *l* of *hl'* is a little doubtful since there is an abrasion obscuring its upper part and it could be an *r*. However, since the *lam auctoris* has a slight curve at its lower end it would seem reasonable to read *l* in this name. There is also a thin scratch which at first sight makes it look like a *t*. However, it will be seen that this scratch continues through the rest of the text and can only be unintentional. As with (i), if this text continued the latter part has been removed by the break in the stone.

(iii) l 'ns bn šddt w s'dh 'l̊y m̊ 'rr ḥy (By 'NS son of ŠDDT: and 'LY helped him against the vengeance of ḤY)

This is an interesting and difficult text which is made the more puzzling by the presence in its second half of several doubtful letters. It is just possible that the *l* of *'ly* is a *z*, though if it is, the cross-stroke is extremely short; the *m* looks more like a *j*, though *m*s of this form are not uncommon; the *'* following the *m* has a small mark inside it and it is possible that it should be read as a *w*, though comparison with the other *'s* and *w* in the text makes *'* the more likely reading. A crack in the stone has obscured the tops of the two *r*s but enough survives to make the reading virtually certain. This crack continues into the *ḥ*, but again the reading seems to be certain. All the names are known: *'ns* has been found three times as a man's name and once as a woman's; *šddt* and *ḥy* are well known.

The interpretation offered above is only one of several possibilities, all of which have their problems. On the whole this seems the safest if possibly the least exciting. *S'dh* would represent Arabic *ša'adahu*, *m* the preposition *min* and *'rr* Arabic *'arâr* (see Lane: 1990/3).

For what they are worth the other two possible interpretations follow.

(a) . . . w s'd h 'l̊ym'rr ḥy (. . . and help. O Most High, against the vengeance of ḤY)

This interpretation would provide a new divine epithet in Safaitic: *'ulyâ*, feminine of *°a'la* and, presumably, the equivalent of the Hebrew *'elyôn*. One should, perhaps, compare the Nabatean names *'bd'ly* and *'bd'l'ly* (Cantineau, *Le Nabatéen*, II, 130–1). Since the epithet is feminine and Allat appears to have been by far the most popular deity among the Safaitic bedu, one may speculate that this was one of her epithets. It is possible that one should take the *h* as the definite article and read the divine name as *h'ly*, in which case it might be suggested that the article was doing duty for the vocative particle as well. However in biblical Hebrew *'elyôn* never seems to occur with the article, so it may be that Safaitic followed the same practice. The syntax would be happier if *s'd* were followed by the pronomial suffix *-h*, but this may have been omitted through haplography. The construction with the imperative preceding the vocative is unusual but has been found before (*CIS* 4961, *SIJ* 836, *MSTJ* 31).

(b) . . . w s'd h'z̊y m̊ 'rr ḥy (. . . and help, H'ZY, against the vengeance of ḤY)

This interpretation requires taking the small smudge at the top of the '*l*' as the cross-stroke of a *z*. This would mean the first occurrence in Safaitic of an invocation to *h'zy*, which is presumably the Safaitic form of the Arabic *al-'uzzâ* and Liḥyanite *hn-'z̊y*. The only previous occurrences in Safaitic of this 'Daughter of Allah' appear to be in the theophoric names *mr'h'zy* (WH 627, 1777, 3820) and *mr''zy* (WH 621) which are probably versions of the same name. For this goddess see the *Encyclopaedia of Islam* (1st ed.) and F. V. Winnett, 'The Daughters of Allah' (*Moslem World*, xxx, 1940, 114–30) and the works cited here. What has been said under possibility (a) about the definite article in the divine name serving also the vocative particle would apply here as well, though of course invocations without the vocative *h* are not unknown (*LP* 286, *NST* 3, *SIJ* 825). The possibility that the pronomial suffix *-h* has been omitted through haplography would also apply to this interpretation. However, given these doubts and the uncertainty of the reading one is hesitant to admit another goddess to the already crowded Safaitic pantheon without firmer evidence.

Finally, it should of course be said that other interpretations of what has been read as *m'rr* are possible, though none have so far been found that are satisfactory.

(iv) l smrn bn jfnẙ h mṣbt (For SMRN son of JFNY is the monument; *or* SMRN son of JFNY [made] the monument)

The *y* of *jfnẙ* is slightly doubtful since there appears to be a loop at the top of the letter, making it look like a *t*. However this loop is probably extraneous to the letter itself. *SMRN* has been found once in Thamudic, *jfnẙ* is new.

The word *mṣbt* has occurred before in C 511 and C 3097, though neither of these texts is any help in deciding its exact meaning. Littman (*Entz* V.191) and the *CIS* have been followed in translating it by 'monument' since in the present state of knowledge it seems impossible to arrive at a more exact translation. If in (iii) either of the interpretations involving a divine name be accepted it would be tempting to give *mṣbt* the sense of *baetyl* or sacred standing stone representing, or dedicated to a deity. It will be remembered that, according to the Islamic writers all three 'Daughters of Allah' had such stones in their sanctuaries and, of course, Arabic *nuṣb* and Hebrew *maṣṣeba* can have this meaning (Winnett, *Daughters of Allah* 114). However, it should be stressed both that these interpretations of (iii) are far from certain and there is no necessary connection between texts (iii) and (iv). Another possibility is, as Littmann suggested (*Entz* 49), that *mṣbt* means some form of enclosure. In view of the drawing on this stone and bearing in mind that it has been impossible to compare the patina on the drawing and the texts, it is possible that we should take *mṣbt* in this sense. However, if this is the case the drawing does not represent the type of enclosure described by Littmann: 'ein kreisförmiger Steinwall, mit einem schmalen nach Osten gerichteten Eingang'.

(v) l 'l̊sw b̊n̊ 'l'm y' .ẙy

The whole text is very faintly written and many of the letters are unclear. The *s* of *'lsw* could be an *ḥ*; the following *bn* is particularly faint; the *l* and *'* of the second name have run into each other; the cross-stroke of the second ' has run into the preceding *y* making it look like an *ṣ*; the letter following this ' is almost totally obscured by an abrasion and could be a' , *t*, *ṣ*, *h* or *y*; the penultimate letter could also be an *ṣ* or an *h*. Given this confusion nothing coherent can be made of this text beyond the first two names and even they are not certain. *'lsw* has not been found before, though *'ls* has; *'l'm* is new. As Winnett and Harding have noted (*WH* 19) Safaitic names are occasionally found written with a final *-w* as in Nabatean and the names they quote which exhibit this feature are also found in Safaitic without the *-w*, as is *'ls.' l'm* also has a Nabatean ring about it, with its reformative *'l-* (see Cantineau, II, 61–2), though this cannot be substantiated. In comparison with Nabatean, Safaitic is relatively poor in names with preformative *'l-*. Of course, the name need not necessarily be Nabatean since it could have come from any Arabic dialect which used *'al* as the definite article rather than *h*.

(vi) l'km bn 'bd

This text is even fainter than the previous one. A horizontal scratch runs between the first name and the *bn*. The second name is particularly difficult to read. Although the *'* is very faint it does seen certain. The *b* of *'bd* is at right-angles to the other *b* in the text and is south Arabian in shape. However, such inconsistencies are occasionally found (see *MSTJ* 20 for a similar example). Both names are attested in Safaitic.

(vii) l qn

Qn is well attested in Safaitic. There is a mark after the *n* and the name may be *qnt*, which is also well attested.

(viii) l qṣyt h ṣlḥ (By QṢYT the honest)

Qṣyt is well attested. The last letter of the text is very faint. The word *ṣâliḥ* or *ṣalîḥ* does not appear to have been found before in Safaitic.

(ix) l 'mr̊t

The *r* is very uncertain since it is obscured by both abrasions and over-scoring. However, given the possibilities presented by what is left of the letter and the attested names composed of the clear letters in the text, *'mr̊t* seems a plausible reading. The fact that the author has written his name inside the drawing may be intended to indicate that he claims it as his own.

(x) l 'zn bn b'd bn b'

The text is written vertically down the left-hand side of the drawing. At the beginning of the text there is a wasm (?) resembling a Safaitic *lḥ* written at right angles to the text. However, it appears to have no relation to the text since the wasm is hammered and the text scratched. A line runs through the middle of the text and many of the letters, particularly the first *bn* have been scratched over, though traces of the original letters are still visible. The second *bn* is particularly faint. All the names are attested in Safaitic, though it should be emphasized that, given the condition of the text, the readings after the first name are highly uncertain.

Abbreviations

CIS — *Corpus Inscriptionum Semiticarum* (1950), V (I), Paris.

Entz — Littman, E. (1901) *Zur Enzifferung der ṣafă-Inschriften*, Leipzig.

Lane — Lane, E. W. (1863–93) *An Arabic-English Lexicon*, 8 vols, London.

LP — Littmann, E. (1943) *Safaitic inscriptions: Publications of the Princeton University Archaeological Expedition to Syria in 1904–05 and 1909*, IV (C), Leiden.

MSTJ — Macdonald, M. C. A. and Harding, G. L. (1976) 'More Safaitic Texts from Jordan', *Annual of the Department of Antiquities of Jordan*, 21, 119–33.

NST Harding, G. L. (1951) 'New Safaitic Texts', *Annual of the Department of Antiquities of Jordan*, 1, 25–9.

SIJ Winnett, F. V. (1957) *Safaitic Inscriptions from Jordan*, Toronto.

TIJ Harding, G. L. and Littmann, E. (1952) *Some Thamudic Inscriptions from the Hashemite Kingdom of the Jordan*, Leiden.

WH Winnett, F. V. and Harding, G. L. (1978) *Inscriptions from Fifty Safaitic Cairns*, Toronto.

SOURCE OF ILLUSTRATIONS

Line Drawings: figures

6 Bender 1968
7 Bender 1968
8 Baly and Tushingham 1971
11 Badawy 1959
18 Harding 1953
23 Betts (unpublished)
26 Betts (unpublished); Perrot 1966
33 Abu Soof 1968; Bossert 1967
34 Mellaart 1970
35 Badawy 1959
46 Kirkbride 1966
47 Ussishkin 1971
48 Mallon and Koeppel 1934, 1940
49 Garstang 1953
50 Mellaart 1970
51 De Vaux 1962
52 Yadin 1958
54 Mellaart 1964
56 Rothenberg
63 Ward 1975
64 Evenari 1971
67 Evenari 1971
68 After Ward 1975
75 Bowen and Albright 1958; Smith 1971
87 Evenari 1971

B7–B9 R. Duckworth
C1 Betts (unpublished)
E1 Köhler (unpublished)

BIBLIOGRAPHY

Abu al-Soof, B. (1968) 'Tell es-Sawwan Excavations (Fourth season)', *Sumer*, 24:3.

Adams, R. M. (1965) *Land behind Baghdad*, Chicago.

et al. (1977) 'Saudi Arabian Archaeological Reconnaissance 1976', *ATLAL*, 1:21.

Amiran, R. (1978) *Early Arad I*, Jerusalem.

(1970) 'The Beginnings of Urbanism in Canaan', in *Near Eastern Archaeology in the 20th Century*, ed. Sanders, New York.

(1969) *Ancient Pottery of the Holy Land from its beginnings in the Neolithic period to the end of the Iron Age*, Jerusalem.

Anati, E. (1968–72) *Rock-Art in Central Arabia*, vols. 1–3, *Expedition Philby-Ryckmanns-Lippen en Arabie*, I (3), Louvin.

(1963) *Palestine before the Hebrews*, London.

Badawy, A. (1959–66) *A History of Egyptian Architecture*, 2 vols. Berkeley.

Baly, D. and Tushingham, A. D. (1971) *Atlas of the biblical world*, New York.

Bell, G. (1907) *Syria, the Desert and the Sown*, London.

Bender, F. (1968) *Geologie von Jordanien* (English edn 1974), Berlin.

Blunt, Lady Anne (1881) *A Pilgrimage to Nejd, the Cradle of the Arabian Race*, London.

Bossert, E. M. (1967) 'Kastri auf Syros', *Archaeologicon Deltion*, 53.

Bowen, R. le B. and Albright, F. P. (1958) *Archaeological Discoveries in South Arabia*, Baltimore.

Braidwood, R. J. and Braidwood, L. S. (1960) *Excavations in the Plain of Antioch*, vol. 1, Chicago.

Breasted, J. H. (1905) *A History of Egypt from the earliest time to the Persian Conquest*, 2nd edn, London.

Burkhardt, F. (1831) *Notes on the Beduin and Wahabys*, London.

Burney, C. (1977) *From Village to Empire,* Oxford.

Butler, H. C. et al. (1919) *Publication of the Princeton University Archaeological Expedition to Syria 1904–5 and 1909: Division I, Geography and Itinerary, Southern Hauran; Division II, Architecture, South Syria,* Leiden.

Butzer, K. W. (1976) *Early Hydraulic Civilization in Egypt,* Chicago.

Cambridge Ancient History, vols. 1–2 (rev. edn), Cambridge.

Childe, V. G. (1952) *New Light on the Most Ancient East* (rev. edn), London.

Crowfoot, J. (1948) in *Archaeological Investigations at Affuleh:* 72–8.

(1937) 'Notes on the flint implements at Jericho', *Liverpool Annals of Archaeology and Anthropology,* 24:35.

De Vaux, R. (1962) 'Les fouilles de Tell el-Far'ah', *Revue Biblique,* 69:212.

Doughty, C. (1888) *Arabia Deserta,* Cambridge.

Dussaud, R. (1929) 'Les Relevés du Capitaine Rees dans le Désert de Syrie', *Syria,* 10:144.

Eaglestone, P. S. (1970) *Dynamic Hydrology,* New York.

Ehrich, R. W. (ed.) (1965) *Chronologies in Old World Archaeology,* Chicago.

Eissfeldt, O. (1966) 'Gabelhürden im Ostjordanland', *Kleine Schriften,* 3:61.

Evenari, M., Shanan, L. and Tadmor, N. (1971) *The Negev,* Cambridge, Mass.

Field, H. (1960) 'North Arabian Desert Archaeological Survey, 1925–1950.' *Papers of the Peabody Museum of Archaeology and Ethnology,* vol. 45, pt. 2.

FitzGerald, G. M. (1934) 'Excavations at Beth-Shan in 1933', *Palestine Exploration Fund Quarterly*:123.

Forbes, R. J. (1955–64) *Studies in Ancient Technology,* 9 vols., Leiden.

Frankfort, H. (1954) *The Art and Architecture of the Ancient Orient,* London.

(1959) 'Town Planning in Ancient Mesopotamia', *Town Planning Review,* 21:99.

(1968) *The Birth of Civilization in the Near East,* London.

Fugman, E. (1958) *Hama: Fouilles et Rechèrches de la Fondation Carlsberg,* 2, 1, Copenhagen.

Gardiner, Sir A. H. (1961) *Egypt of the Pharaohs,* Oxford.

Garrard, A. N. et al. (1977) 'A Survey of prehistoric Sites in the Azraq Basin', *Paleorient*, 3:109.

Garstang, J. (1953) *Prehistoric Mersin*, Oxford.

Glueck, N. (1959) *Rivers in the Desert*, London.

(1951) 'Explorations in Eastern Palestine', 4 (1–2), *Annual of the American Schools of Oriental Research*, 25–8:428.

Harding, G. L. (1959) *The Antiquities of Jordan*, London.

(1953) 'The Cairn of Hani', *Annual of the Department of Antiquities of Jordan*, 2:8.

Heinrich, E. (1958) 'Die "Inselarchitektur" des Mittelmeergebietes und ihre Beziehungen zur Antike', *Archaeologischer Anzeiger*:89.

Helbaek, H. (1972) 'Samarran irrigation agriculture at Choga Mami in Iraq', *Iraq*, 31:35.

Helms, S. W. (1973–7) 'Reports on Jawa', in *Levant* 5:127; 7·20; 8:1; 9:21.

(1976) *Urban Fortifications of Palestine during the Third Millennium BC* (unpublished dissertation, Institute of Archaeology, London).

Hennessy, J. B. (1969) 'Preliminary report on a first season of excavations at Teleilat Ghassul', *Levant*, 1:1.

(1967) *The Foreign Relations of Palestine during the Early Bronze Age*, London.

Ionides, M. G. (1937) *The Regime of the Rivers Tigris and Euphrates*, London.

Jacobsen, T. and Adams, R. M. (1958) 'Salt and Silt in Ancient Mesopotamian Agriculture', *Science*, 25:153.

Kaplan, J. (1969) 'Ein el Jarba. Chalcolithic remains in the Plain of Esdraelon', *Bulletin of the American Schools of Oriental Research*, 194:2.

Kenyon, K. M. (1979) *Archaeology in the Holy Land*, 4th edn, London.

(1960–5) *Excavations at Jericho*, 2 vols., London.

Kirkbride, D. (1966) 'Five Seasons at the Pre-Pottery Neolithic Village of Beidha in Jordan', *Palestine Exploration Quarterly*:8.

Kraeling, C. H. and Adams, R. M. (eds) (1958) *City Invincible: A symposium on Urbanization and Cultural Development in the Ancient Near East. Held at the Oriental Institute of the University of Chicago, 4–7 December, 1958*, Chicago.

Lampl, P. (1968) *Cities and Planning in the Ancient Near East*,

London.

Lapp, P. (1970) 'Palestine in the Early Bronze Age' in *Near Eastern Archaeology in the 20th Century*, ed. Sanders, New York.

Maitland, R. A. (1927) 'The Works of the "Old Men" in Arabia', *Antiquity*, 1:197.

Mallon, A. and Koeppel, R. (1934, 1940) *Teleilat Ghassul*, 2 vols.

Mayerson, P. (1960) *The Ancient Agricultural Regime of Nessana and the Central Negeb*, London.

Mellaart, J. (1975) *The Neolithic of the Near East*, London.

(1970) *Excavations at Hacilar*, 2 vols., Edinburgh.

(1966) *The Chalcolithic and Early Bronze Ages in the Near East and Anatolia*, Beirut.

(1964) 'Excavations at Çatal Hüyük', *Anatolian Studies*, 14:8.

(1962) 'Preliminary Report on the Archaeological Survey in the Yarmuk and Jordan Valley', *Annual of the Department of Antiquities of Jordan*, 6–7:126.

Meshel, Z. (1974) 'New Data about the "Desert Kites"', *Tel Aviv*, 1:129.

Moore, A. M. T. (1973) 'The Late Neolithic in Palestine', *Levant*, 5:36.

Mountfort, G. (1966) *Portrait of a Desert*, London.

Mowry, L. (1953) 'A Greek Inscription from Jathum', *Bulletin of the American Schools of Oriental Research*, 132:34.

Mumford, L. (1961) *The City in History*, London.

Musil, A. (1928) *The Manners and Customs of the Rwala Beduins*, New York.

Neuville, R. (1934) in *Mallon* 1934:55–65.

Neuville, R. and Mallon, A. (1931) 'Le début de l'âge des métaux dans les grottes du désert de Judée', *Syria*, 12:30–32.

Oates, D., Oates, J. (1976) 'Early irrigation agriculture in Mesopotamia', in *Problems in Economic and Social Archaeology*, ed. Sieveking, London:109.

Perrot, J. (1968) *Préhistoire Palestinienne: Extrait du supplément du Dictionnaire de la Bible*, 7:286–446, Paris.

(1966) 'La Troisième Campagne de Fouilles à Munhata', *Syria*, 43:49.

(1964) 'Les Deux Premières Campagnes de Fouilles à Munhatta', *Syria*, 41:323.

(1955) 'The Excavations at Tell Abu Matar, near Beersheba', *Israel*

Exploration Journal, 5:73.
Poidebard, A. (1934) *La Trace de Rome dans le Désert de Syrie*, Paris.
(1928) 'Reconnaissance aérienne au Ledja et au Safa', *Syria*, 9:114.
Rees, L. W. B. (1929) 'The Trans-Jordanian Desert', *Antiquity*, 3:389.
Renfrew, C., Dixon, J. E. and Cann, J. R. (1966) 'Obsidian and early cultural contact in the Near East', *Proceedings of the Prehistoric Society*, 32:30.
Rothenberg, B. (1972) *Timna, Valley of the Biblical Copper Mines*, London.
(1970) 'An archaeological survey of South Sinai', *Palestine Exploration Quarterly*: 2.
Schmidt, J. (1963) *Die agglutinisierende Bauweise im Zwei-Stromland und in Syrien*, Berlin.
Smith, N. (1971) *A History of Dams*, New Jersey.
Stager, L. E. (1976) 'Farming in the Judean Desert during the Iron Age', *Bulletin of the American Schools of Oriental Research*, 221:145.
Stein, Sir A. (1940) 'Surveys of the Roman Fontier in Iraq and Transjordan', *Geographical Journal*, 195:428.
Thesiger, W. (1959) *Arabian Sands*, London.
Ucko, P. J. (ed.) (1970) *Man, Settlement and Urbanism: Research Seminar in Archaeology and Related Subjects*, London.
(1969) *The Domestication and Exploitation of Plants and Animals*, London.
Ussishkin, D. (1971) 'The Ghassulian Temple in Ein Gedi and the Origin of the Hoard from Nahal Mishmar', *Biblical Archaeologist*, 34:23.
Waechter, J. D'A. and Seton-Williams, V. M. (1938) 'Excavations in Wadi Dhobai, 1937–38 and the Dhobaian Industry', *Journal of the Palestine Oriental Society*, 18:172 and 292.
Ward, R. C. (1975) *Principles of Hydrology*, London.
Ward, W. A. (1969) 'The Supposed Asiatic Campaign of Narmer', *Mélanges de l'Université Saint-Joseph de Beyrouth*, 45:296.
Winnett, F. V. (1973) 'The Revolt of Damasi', *Bulletin of the American Schools of Oriental Research*, 211:54.
(1951) 'An epigraphical expedition to North-Eastern Transjordan', *Bulletin of the American Schools of Oriental Research*, 122:49.
Yadin, Y. (1963) *The Art of Warfare in Biblical Lands*, London.

(1958) 'Solomon's City Wall and the Gate at Gezer', *Israel Exploration Journal*, 8:80.

(1955) 'The Earliest Record of Egypt's Military Penetration into Asia?', *Israel Exploration Quarterly*, 5:5.